Through the Eyes of John Muir

PRACTICES IN ENVIRONMENTAL STEWARDSHIP

Meets Common Core Standards for Third and Fourth Grades

by Janice Kelley

Other books by the Author
    Mornings on Fair Oaks Bridge, Watching Wildlife at the Lower American River
    In Nature's Time
    What Song does the Rain Sing? A Creative Writing and Exploration Journal for kids
    *Inspired by John Muir, A Guide to Studying Nature, Capturing our Stories and Advocating for Wild Places
        (the complementary journal to teacher's guide)

Copyright © 2016 by Janice Kelley. All Rights Reserved.

Content of this book has been edited from the original publication Field Trip Curriculum for the John Muir National Historic Site; written by Janice Kelley, copyright 2013, published by California State University, Sacramento.

Assignment sheets, interview questions and other educational materials may be duplicated as needed for educational purposes only.   Meets Common Core Standards for Grades 3 and 4

Cover photos Sequoia National Park. Photo credit: Janice Kelley

Through the Eyes of John Muir/ Janice Kelley. —2nd edition. ISBN: 978-0-9715467-8-3

Acknowledgements

I owe many thanks and sincere gratitude to my "support team" who informed and guided me through the research and development process of this project. The opportunity to draw from the extraordinary life of John Muir blended both my passion for exploring the natural world and my skills and experiences as an interpreter, writer and storyteller.

Dr. Elizabeth Erickson and Dr. David Rolloff as my first readers who shared their enthusiasm for the life of John Muir and the lessons we learn from him. Matthew Holmes, who shared his vision for education programs representing the interpretation and education team at the John Muir Historic Site.

My son Daniel, who sacrificed many "mom together" hours as I researched, wrote, typed and revised my original field trip curriculum project and its transformation into this book.

March 2016

"One learns that the world, though made is still being made.
That this is still the morning of creation."

JOHN MUIR

Introduction

JOHN MUIR is often remembered as an environmental activist, naturalist and prolific writer. His timeless quotes are written in books and posted to websites, displayed on banners that hang in national park visitor centers, and spoken in videos about Muir's life, national parks, wild lands and stories of the environment.

His enduring legacy of environmental stewardship remains strong more than 100 years after his death.

Yet, this essence of Muir represents only a fraction of his life. He was a man of many talents, trades and interests. Muir was especially skilled at influencing the influencers - his relationship with President Theodore Roosevelt led to the preservation of millions of acres of forest and wild lands.

Muir lived his last 25 years with his family on a large fruit ranch in Martinez, California. He wrote many books and articles in the "Scribble Den." The ranch, now the John Muir National Historic Site, is located about 35 miles northeast of San Francisco.

His life at the Strentzel-Muir Ranch parallels the development of early California agriculture. It is through Muir's legacy that we can learn about immigrant labor, efficient crop production, water conservation, product marketing and transportation.

John Muir and his wife Louise (Louie) were devoted to their family and raised two daughters at the ranch. Financial gains from the lease and sale of parts of this property supported Muir during the years he was an activist.

He was born in Scotland, coming to America as a young boy with his family. His history is a reflection of many immigrants who turned to farming the land to make their living.

Muir also witnessed the urbanization of Alhambra Valley in the early 1900s during the final years of his life. This urbanization was in part a result of how the Industrial Revolution transformed American landscapes and places of work.

Inquiry-Based Curriculum

This guide features a series of lessons that cross the disciplines of language arts, social studies and science.
- Gather evidence from field studies to solve problems
- Conduct research to determine how people, communities and professions change over time to keep up with the changing dynamics of the world.
- Learn "how to think," instead of "what to think."
- Study the relationships to Muir's life and actions to their own life and community.
- Learn and apply personal responsibility and develop a personal vision for the future.
- As Muir used the power of his writing to persuade, students will have opportunities to write and share a persuasive story.

Students can hold Muir as their role model for environmental stewardship, scientific research and practice. He represents individuals around the world – in history and contemporary times - who struggle and persevere to make the world a better place for all. Through Muir's personal reflections, students learn his values. They read what he thought and wrote about the natural world. Students discover the actions he took to make a difference in his community and the world.

All lessons meet Common Core Standards and some also align with National Science Standards.

Through the Eyes of John Muir

Curriculum Guide

- John Muir quotations
- Field investigations and games
- Early California agriculture
- Personal and community histories
- Environmental conservation and stewardship

Contents

Introduction to the Curriculum Guide 5

Purpose of the Curriculum Guide 7

Chapter One: Orientation to John Muir 9

Summary of Orientation to John Muir 10

Introducing: Who was John Muir? 11

Reading Reflections Unit 12

Lesson One: John Muir Biography 12

    Study Questions 14

Lesson Two: John Muir Quotations Reflection Project 17

    Assignment Sheet 18

    Discussion Sample 19

    List of Quotations 20

Lesson Three: John Muir Timeline 22

    Timeline 23

Puzzles and More 24

    Story of John Muir Worksheets 25

    Who was John Muir? Crossword puzzle 28

    John Muir: The Scientist Crossword Puzzle 30

    Science Crossword puzzle 32

Agriculture 34

The Art of Agriculture 35

    Brief History of Fruit Crate Labels 37

Student Bibliography 39

Resources for Teachers 40

Chapter Two: Field Studies 42

Nature Scene Investigations (NSI) 43

Meet a Tree 44

    Tree Study Questions 46

Explore a Special Place 47

    Special Place Questions 49

Creek Study 50

    Creek Study Questions 52

Nature Games 53

Recipe for a Healthy Forest 53

What am I? Guessing game 56

Chapter Three: Classroom Projects 61

Summary of Classroom Projects 62

History and Language Arts Unit 63

Lesson One: Who am I? 64

Lesson Two: People of the past 66

    1880s Child 66

        1880s Child: What do you have in common? 69

    Family History Project 70

        Family History Assignment Sheet & Interview Questions 71

Lesson Three: Other People, Other Places 72

    Storytelling Decision Map 75

    Folklore, Legends, Myths and more 76

Lesson Four: Our Changing Community: Community Photo Album Project 78

    Community Photo Album Assignment Sheet 80

    The Changing Alhambra Valley 81

Science, Conservation and Advocacy Unit 83

Lesson Five: Meet a Modern Day John Muir 83

    Poster Assignment Sheet 85

Lesson Six: My Role as a Community Steward 86

Lesson Seven: Agency Roles to Protect Wild lands And Wildlife 88

    Agency Role Assignment Sheet 91

Lesson Eight: My Role as a Community Advocate 92

    Advocacy Assignment Sheet 94

Lesson Nine: My Vision for the Future 95

Glossary 99

Curriculum Guide Bibliography 101

Introduction to the Curriculum Guide

This guide presents nearly three dozen lessons covering a broad spectrum of topics: various aspects of John Muir's life, California agricultural and immigrant history, urbanization of farmland, nature-based investigations, storytelling, what makes a healthy forest and more. It is important to pose the question, Why is John Muir singularly important to study within the context of history, social studies, studies and science? What makes him the connecting point in this curriculum that few others can do?

As a leading authority in his day, Muir understood and applied the connections between people, wildlife, the earth and the ecosystems that support life (in the days before the word "ecosystem" was officially used). He was deeply connected to each of the subjects addressed in this guide. His role as a rancher, scientist, writer and activist, are only some of his many talents and professions. As one person, through tireless and steadfast effort, he contributed to significant and positive changes in the United States and around the world. His influence, ideology and legacy of environmental activism continues on through the lives of countless others today in the United States and abroad more than 100 years after his death.

"Nothing goes unrecorded. Every word of leaf and snowflake and particle of dew ... as well as earthquake and avalanche, is written down in Nature's book."

By learning about John Muir, students are likely to find themselves: as immigrants, gardeners, explorers or wanderers, as future inventors, entrepreneurs, scientists, naturalists, writers and/or activists. Students can hold John Muir as a role model to guide their own actions. A selection of Muir's quotations are included in this guide, in addition to a link to the National Park Service website where teachers and students can read hundreds of other quotes.

John Muir as an environmental activist was regularly in the center of controversy over difficult choices regarding the environment. Debates continue today over water resources issues, the logging of forests, the destruction of native habitats and the degradation of air and water quality. The lessons in this guide give students the opportunity to launch into exploring, researching and discussing the types of issues. Students will undoubtedly find themselves immersed in addressing these challenges of today and those that will be of concern in the predictable future.

Through a study of Muir's life, students can see that each of them can make a difference in their world. What students learn (and all those who study John Muir) is even more important than identifying his many accomplishments. Students can begin to understand they have both the opportunity and the power to create the world they want to live in, to choose what is acceptable, and to take the actions for which they will be held accountable. Students can witness by studying the life of John Muir, and follow in his footsteps, that amazing things can happen to transform their communities and, in turn, the world into a better place for all.

On Developing an Environmental Ethic

Although the scope of this guide is far more than environmental education alone, a significant part of John Muir's life was as a champion for wild places. In respect for Muir's legacy, it is important to consider that a modern study of environmental awareness and issues has become increasingly more complex and multi-layered.

Environmental messages are deeply rooted and influenced by individual experience, geography and culture.[1] These three factors contribute to the formation of a student's environmental ethic - their behavior and actions in relationship to their environment. The practice of ethical behaviors is a common thread repeated throughout the study of John Muir, although the word ethics is not used specifically. As a consequence of the widely diverse cultural backgrounds of the student population in California classrooms - in native languages, cultural values and beliefs - creating the foundation for developing a shared environmental ethic that includes behaviors to promote long-term sustainability for the good of all people may be a challenge if it runs contrary to deep-seated cultural beliefs of the family.

Coupled with the long held cultural beliefs, is the widely documented notion of a student's increasing disconnection from direct experience with the natural world. The unfortunate consequence of this disconnection is that the information students (and many adults) learn about the environment is not the result of direct experience. Information tends to be mediated by social institutions and social values.[2]

> "Environmental education can help foster greater understanding and appreciation of the environment. Maybe even more important is providing students with basic knowledge and experiences, a core of information to build up and use throughout their lives as they make choices and decisions that affect the environment...Teach them how to think about environmental issues, and not what to think about them."[3]

[1,2] Corbett, J.B. (2006). Communicating Nature, How we create and understand environmental messages. Washington, DC: Island Press.
[3] Brown, K. (1998). Environmental Service-Learning. St. Louis, MN: Tree Trust.

Purpose of the Curriculum Guide

This curriculum guide provides a series of interdisciplinary activities to learn about John Muir, field studies to conduct at a nature area as an outdoor classroom, and concludes with lessons and practices to help students became stewards of the environment and "champions" in their own community.

Classroom and field assignments require students to read, research, compare, contrast, synthesize, problem solve, work independently and collaboratively. Learning activities apply both oral presentation and writing skills. All lessons in this guide can stand-alone, so teachers may select as few or as many as desired.

Throughout this experience, students participate in a variety of learning opportunities to develop and refine the critical skills they need to be successful in school and make a positive contribution to their communities. Students can engage in activities that prompt individual and collective reflection on controversial topics that stir debate often based on ideology and personal backgrounds. They will learn to look at the HOW of an issue in addition to the WHAT of an issue. These skills can help students engage in thoughtful debates that lead to informed decisions about a shared future on earth.

Arrangement of the Guide

This guide is divided into three sections: 1) Orientation to John Muir, 2) Field Studies and 3) Classroom Projects. The content is arranged to be introduced in order, as students become more familiar with the life of John Muir and increasingly more aware of his or her own roles and responsibilities within the community where he/she lives.

The guide features many different lessons with multiple activities within each lesson, in each of the three sections. A summary of lessons is included at the beginning of each section. These lessons can help students understand the life and work of John Muir with the intention that they will continue in his footsteps to make a positive difference in their respective communities and work to conserve wild places for the benefit of all.

Orientation to John Muir. Orients teachers and students to John Muir and the historic site. Materials include classroom activities and a suggested reading list for students describing John Muir's life. A list of YouTube videos and storyteller recordings of

John Muir's adventures are also included in the student bibliography. Teachers will find a comprehensive list of print, web and organization resources to supplement opportunities for student learning.

Field studies. Presents a series of thematic lessons to be conducted outdoors that create relevant connections for the students to John Muir's actions and beliefs, and standard practices of his era.

Classroom Projects. Describes suggested activities and supplementary materials on historical, scientific and environmental topics for student research and discussion after field study experiences. Includes opportunities to involve guest speakers, engage the community and participate in real community problem solving.

John Muir wrote and illustrated extensively of his travels and adventures in a series of journals. Classroom projects involve a significant amount of writing.

Recommendations for Students: To continue in the spirit of Muir's journal keeping, it is highly recommended that students keep all their work assembled as an ongoing journal in a single 3- ring binder.

- Assists the student keep track of the projects in progress;
- Provides a meaningful student reference upon completion; and simplifies evaluation of student achievement and skill advancement.

John Muir brings a diverse range of themes to be studied. You may consider expanding on lessons of conservation and responsibility to stimulate ideas for creating a "History Day" around his life and accomplishments.

Visit the Sierra Club website to find information about upcoming events, photographs, writing, video, quotes, additional lessons and biographical information about John Muir. http://vault.sierraclub.org/john_muir_exhibit/

Orientation to John Muir

Chapter One

Chapter One Summary

Background

This section provides the opportunity to orient students to John Muir. Students acquire and demonstrate a basic level of knowledge by completing a series of reading and writing activities. This section focuses on the value of Muir's lifelong contributions to the conservation movement.

Lessons in this Section

Reading Reflections

- John Muir Biography
- John Muir Quotations Reflections Project
- John Muir Timeline and discussion

Puzzles and More

- Story of John Muir
- "Who was John Muir?" Crossword Puzzles
- John Muir: The Scientist
- Science

Art of agriculture

- Fruit Crate Labels

Who Was John Muir? — Chapter One

Background

John Muir's life was shaped by multiple influences and events that fed his passion for conserving wild places in America and around the world. His life is best understood in three distinct phases: as a child in Scotland through his early adulthood as a factory worker, his call to the natural world when he ventured on a 1,000 mile walk, and his arrival at Yosemite, and his final years as a rancher and activist.

Goals

- Read, view and discuss significant moments and influences in the life of John Muir that laid the foundation for his passion for wild places and his work as an activist.
- Witness that a single person can make a world of difference.

Objectives

By the end of this unit, students will be able to:

1. Identify three major influences or events that shaped the life of John Muir.
2. Discuss three or more reasons why John Muir's life is and continues to be nationally recognized nearly 100 years after his death.
3. Cite at least one quotation of John Muir and explain how it relates to the student's life.

Assignments

1. The **Reading Reflections Unit** consists of reading, viewing, listening, discussion, writing and art activities.

2. **Puzzles and More** includes activities that create an opportunity to focus on specific aspects of John Muir's life and work.

Reading Reflections Unit   Chapter One

Background

This unit presents a set of activities that orient students to the life of John Muir and the key influences in his life that shaped his actions.

1. **John Muir Biography** is a reading activity paired with student study questions and class discussion.
2. **John Muir Reading Reflection Project** uses his quotations as a basis for students to understand his passion and concern about the natural world.
3. **Timeline** is an opportunity to highlight important moments in the life of John Muir with important moments and activities in the student's life.

Lesson One: John Muir Biography   Ch. 1 Lesson One

Background

Many authors before and since John Muir's death have written about Muir. Since each author expresses a different aspect of this complex man, reviewing multiple selections provide even more insight into Muir's values and philosophy.

Objectives

1. Complete student study questions describing biographical information about John Muir.
2. Actively participate in class discussion for reading, movies or audio recordings.

Procedures

1. Assign students to read either or John Muir, America's First Environmentalist by Kathryn Lasky or John Muir, Protecting and Preserving the Environment by Henry Elliot. Students may read alone or with a parent.
2. Listen to audio recording of John Muir and Stickeen, as told by Garth Gilchrist.
3. Watch a YouTube.com video during class. Encourage students to watch others at home with family members.
4. .Some selections are written for a high grade reading level and may require a parent's assistance with reading. The parent may aloud to the student.
5. Encourage students to share reading materials with parents so they can begin discussions at home in the following areas: a) the need for and importance of conservation and b) how students can become involved in their own community.

References

Literature

Elliot, H. (2009). John Muir, Preserving and Protecting the Environment. Crabtree Publishing Company: New York, NY.

Lasky, K. (2006). John Muir, America's First Environmentalist. Candlewick Press: Cambridge, MA.

Audio recordings

Gilchrist, G. (2000). My life of adventures. John Muir. On CD. Dawn Publications: Nevada City, CA

Gilchrist, G. (2000). Stickeen and John Muir's other animal adventures. On CD. Dawn Publications: Nevada City, CA.

*Videos

Biography of John Muir, No. 1 "A glorious journey" 10 min. view at
http://www.YouTube.com/watch?v=-CDzhIvugw8

Common Core Standards

Reading Standards for Literature, Third grade. 1,3,4,7   Fourth grade. 1,3,4,7

John Muir Biography Study Questions   Ch.1   Lesson One

Respond to the following questions with two or three complete and detailed sentences.

1. What was one of John Muir's first encounters with nature as a young boy?
2. Describe one or more of Muir's inventions.
3. List three places in the United States that he visited.
4. What was the reason he went on his 1,000-mile walk to the Gulf of Mexico? Describe one or more of the experiences he shared about his journey.
5. What did Muir discover about how Yosemite was formed?
6. What else did Muir do in Yosemite?
7. Describe one or more of his "high adventure" activities
8. Why was John Muir living on a ranch?
9. Describe what motivated Muir to spend his life protecting the wilderness.
10. Did John Muir act alone or did he have support to make changes happen? If so, name two people that helped John Muir make changes.
11. What were the subjects that Muir wrote about in his articles, books and journals?

John Muir Biography Study Questions Answer Sheet

What was one of John Muir's first encounters with nature as a young boy?

He was with his brother searching tide pools for crabs and other interesting creatures of the sea near his childhood home in Scotland. He climbed the walls of the ruins of Dunbar Castle. Muir regularly visited the shore and the meadows where he could study the creatures and wildlife living there.

Describe at least one of John Muir's inventions.
1. Study desk
2. Sawmill was used in a stream to cut wood
3. Barometers

What was the reason he went on his 1,000-mile walk to the Gulf of Mexico? Describe one or more of the experiences he shared about his journey.

John Muir knew he could not work in a factory again because there was so much to explore and learn about the outdoors. For the first time in his life, he saw tall trees and dense forests. He saw high mountains and deep valleys. Muir took note of flowering plants. There were kind families that offered him food and shelter.

List three places in the United States that Muir visited.

Yosemite National Park	San Francisco
Alaska	New York
Mt. Whitney	Niagara Falls
Lake Superior	Kings Canyon

What did he discover about how Yosemite was formed?

Glaciers had cut their way through the rocks to carve the Yosemite Valley. He measured the height and width of rocks and boulders as evidence. Many people disagreed with him at the time, although later he was proven to be right.

What else did John Muir do in Yosemite Valley?

John Muir was a nature guide, leading visitors to his favorite scenic spots. As a guide, he pointed out the evidence of rocks carved by glaciers in several locations. He was also a shepherd. He also met President Roosevelt to camp in Yosemite to convince him how important it was to establish Yosemite as a national park for the public good.

Name some of his "high adventure" activities.

1. John Muir climbed Mt. Whitney when there was a severe snowstorm and it was extremely cold. His body was frozen and he decided not to keep climbing in that weather.
2. He climbed Mt. Shasta. During his climb the weather changed from sun to a heavy snowfall.
3. A third high adventure activity was when he climbed to the top of a Douglas fir during a windstorm to listen to the needles sing in the wind.

Why did John Muir live on a ranch?

John Muir married John Strentzel's daughter. Strentzel owned a large fruit ranch in Martinez, CA. When Muir married Louie, they lived on the ranch and raised their family. John managed the ranch after his father-in-law died.

What motivated John Muir to spend his life protecting wild lands?

He believed that the wonders of nature in America should be preserved as national parks for people to visit and enjoy. He loved the wilderness and finding out more about nature. When Muir saw indications that man was destroying natural areas, he decided to share his experiences with others by writing articles and books.

Did John Muir act alone or did he have support to make changes happen? If so, name two people who supported him?

No, John Muir did not act alone most of the time. He had support from Ezra and Janine Carr, John Swett and Teddy Roosevelt. Muir helped to form the Sierra Club. By working together, Congress passed laws that protected the wilderness and created national parks for people to enjoy, instead of being destroyed by logging.

List some of the topics that Muir wrote about.

Glaciers	Trips to Alaska
His first summer in the Sierra	Stickeen
Saving wilderness	Preserving Yosemite as a National Park
Mountains of California	

John Muir Quotations Reflection Project   Ch 1 Lesson Two

Background

Hundreds of quotes from John Muir are printed in books, magazine articles, websites and numerous collateral materials. The topics are vast – ranging from political challenges to the awe he experienced from the forces of nature. The quotations that follow are arranged within seven guiding principles (as interpreted by the author) to frame Muir's perspective on the world.

Objectives

1. Discuss the relevancy and meaning of John Muir's basic values of the natural world as expressed through his quotations.
2. Create a tangible product, such as a poem, collage, journal illustration, photograph or short story that describes the essence of the quotation's meaning.

Procedures

1. Identify a quote from the list provided in this guide to discuss meanings with class. This project can be repeated weekly.
2. See the following discussion example.
3. Allow 20-30 minutes to begin in class and complete at home to return on Friday.
4. Distribute the John Muir Quotations Reflection handout that explains the project to take home. Invite them to share the assignment sheet with their parent(s) to review in case the student needs assistance for one or more of the projects.
5. Ask students to write original quotation somewhere on their project as reference.
6. Students share their work the day they bring it back to school to post in classroom.
7. Encourage the student to choose different type of artistic media each week.

Common Core Standards

Third grade. Reading Standards for Literature. 1,3,4,6,7
Fourth grade. Reading Standards for Literature. 1.3.4

John Muir Quotations Reflection Assignment Sheet

Our class will be studying a new quotation by John Muir every week. Following a class discussion, each student will complete a project that illustrates what that quote means to you. You will be given limited class time to think about and begin your project.

Instructions

1. Please write the what subject you think the quotation describes, such as "beauty," "animals," or "nature."
2. The assignment will be due on Friday of the same week.
3. Please create any one of the following projects.

Projects

1. Poem
2. Collage (words, pictures or objects)
3. Photograph postcard with a caption - the quotation (or first sentence) must also be written on the bottom edge of your project
4. Illustration or paragraphs written as a journal page
5. A different creative or new media technology project with teacher approval

John Muir Quotations Reflection Project - Discussion Sample

Procedures

Write the quotation on the whiteboard or project with a screen at specific intervals, such as weekly. Ask a student to read the quote you have written with feeling as if John Muir were speaking it.

"These beautiful days must enrich all my life. They do not exist as mere pictures...but they saturate themselves into every part of my body and live always."

Sample Discussion Questions

1. What do you think John Muir was talking about in this quote? Is a beautiful day something that exists in nature?
2. For those of you who have seen a waterfall or sunset in person, do you think seeing a photograph gives you the same experience as seeing it in person? What do you think you will remember most?
3. What sensory experiences do you miss in a photograph that you get in person? (e.g. the crash of a waterfall, the cool mist blowing on your skin. Or, the changing colors of the sky and feel of the air getting cooler as the sun goes down)
4. What John Muir appears to be telling us through his words is that he thinks the natural world is so beautiful and so special that the experience gets inside of us. Seeing a photograph is not enough. He needed to be there in person to get the fullest possible experience of nature he could get.
5. He realized his life was better because of being in nature. So, he cared a lot about it.
6. In just this one quotation, we find out why John Muir was so interested in caring for the natural world. The natural world became part of his body to nourish and heal it.
7. We need to ask ourselves, do we think our relationship with the beauty of nature makes our life better?
8. Are we content with seeing pictures of nature or do we want to experience it in person?
9. Do we want to care for nature so it remains beautiful? If so, what actions can we take to care for nature?

John Muir Quotations

The phrase in **boldface** type represents the guiding principle as interpreted by the author.

Respect for all creatures and creation.

"How narrow we selfish, conceited creatures are in our sympathies! How blind to the rights of all the rest of creation!"

"How many mouths Nature has to fill, how may neighbors we have, how little we know about them, and seldom we get in each other's way!"

John Muir's passion and connection to the natural world.

"God himself seems to be always doing his best here, working like a mean in a glow of enthusiasm."

"In every walk with Nature, one receives far more than he seeks."

All creation is connected, beautiful and whole.

"One learns that the world, though made, is yet being made. That this is still the morning of creation."

"Nothing goes unrecorded. Every word of leaf and snowflake and particle of dew ... as well as earthquake and avalanche, is written down in Nature's book."

"When we try to pick out anything by itself, we find it hitched to everything else in the universe."

The drama of the real experience does not match reading about the outdoor world or seeing pictures.

"Then it seemed to me the Sierra should be called, not the Nevada, or Snowy Ridge, but the Range of Light."

"These beautiful days must enrich all my life. They do not exist as mere pictures . . . but they saturate themselves into every part of the body and live always."

Muir Woods, photo by Janice Kelley

Nature is a good mother for nourishment and healing.

"Camp out among the grass and gentians of glacier meadows, in craggy garden nooks full of Nature's darlings. Climb the mountains and get their good tidings. Nature's peace will flow into you as sunshine flows into trees. The winds will below their own freshness into you, and the storms their energy, while cares will drop off like autumn leaves."

"Nature is a good mother, and sees well to the clothing of her many bairns—birds with smoothly imbricated feathers, beetles with shining jackets, and bears with shaggy furs."

"Everyone needs beauty as well as bread places to play in and pray in, where nature may heal and give strength to body and soul alike."

Sequoia National Park, photo by Janice Kelley

Human and ecological destruction is driven by the relentless push for profit.

"I often wonder what man will do with the mountains—that is, with their utilizable, destructible garments. Will he cut down all the trees to make ships and houses? If so, what will be the final and far upshot? Will human destructions like those of Nature—fire and flood and avalanche—work out a higher good, a finer beauty? ... What is the human part of the mountains' destiny?"

Sequoia National Park, photo by Janice Kelley

Nature is an everlasting chain.

"This grand show is eternal. It is always sunrise somewhere; the dew is never all dried at once; a show is forever falling; vapor is ever rising. Eternal sunrise, eternal sunset, eternal dawn and gloaming on sea and continents and islands, each in its turn, so the round earth rolls."

John Muir Timeline

Ch 1. Lesson Three

Background

A series of events and people shaped the beliefs and direction of John Muir's life. His life is divided into three phases: first, childhood to young adult as a factory worker, second, wilderness explorer and third, fruit rancher, father and environmental activist.

Goal

Stimulate class discussion and connect significant events in John Muir's life listed in the timeline to student experiences.

Discussion Questions

- Ask students to list important markers in their life so they can begin to create their own timeline.
- In what ways did John Muir show perseverance - never giving up? Give an example of when a student has persevered and not given up.
- John Muir was recognized for doing a lot of great things, including establishing the Sierra Club to protect the wilderness and wildlife. What contributions has the student made to their classroom, their family, and their school? Who do they know that contributes something to their school? (e.g. a librarian, teacher or the janitor)
- John Muir did not act alone. He created support for himself and relied on mentors. Who does the student go to as a source of support or as a mentor?
- Muir lost the battle to save Hetch Hetchy and it was his biggest defeat. Ask students if they have worked heard to achieve something very important and did not make it?
- Once Muir had a family, he realized more than ever that preserving the wilderness was even more important for the future generations beyond his lifetime. What makes it important for you to be thinking about the future of the wilderness?
- Stickeen was a special dog. Muir also developed relationships with squirrels. Does the student have a special relationship with an animal? If so, what activities does the student and pet do together?

Extended learning opportunity

- Map John Muir's travels across the United States as a class project.

Common Core Standards

Third grade. Speaking and Listening. 1.4.6
Grade Four. Speaking and Listening 1.4.6

Timeline - Selected Highlights of John Muir's Life

Year	Event
1838	Birth
1849	Immigrates to America (Wisconsin)
1860	Enters inventions into the Wisconsin State Fair
1860	Meets Jeanne and Ezra Carr at the university
1861	Enrolls in Wisconsin State University
1867	Machine accident blinds John Muir temporarily
1867	1,000-mile walk
1868	Arrives in Yosemite
1874	Publishes first article regarding conservation practices
1880	Marries Louise Strentzel
1880	Travels to Alaska with Stickeen
1881	First child is born
1891	Retires from farming and becomes an environmental advocate
1892	Forms Sierra Club
1903	Camps with President Theodore Roosevelt in Yosemite Valley
1907	Hetch Hetchy Dam fight begins
1914	John Muir dies of pneumonia

For additional information and dates to add to the timeline, see http://www.sierraclub.org/john_muir_exhibit/life/chronology.aspx

Puzzles and More

Background

Each of the following activities focuses on a different aspect of the life of John Muir. The science crosswords contain words that students may not recognize. To create a more meaningful learning experience, present the answer sheet to students and ask them to research and write the meaning of the words. Next, give students the blank puzzle, so they use the meanings they found, and match them to puzzle clues to complete the work.

- Story of John Muir worksheet
- Who was John Muir Crossword Puzzle
- John Muir the Scientist Crossword Puzzle
- Science Crossword Puzzle

Common Core Standards

Third Grade. Keys Ideas and Details.1. Craft and Structure, 4
Fourth Grade. Keys Ideas and Details.1. Craft and Structure, 4

Story of John Muir

Complete the sentences below to read the story by using words from the box. Words can only be used once.

	Key Words	
respect	wandered	observe
treasure	active	exploring
ancient	magnificent	Risks
examine	struggle	defend
ambition	reverse	triumph
nation	surveyed	Globe
ordeal	opponent	outcome
individuals	journey	solutions

John Muir learned to love and _____ nature when he was a young boy at his family's farm in Scotland. He _____ through fields and forests to _____ the birds, squirrels and other wildlife. Muir always _____ nature. "Climb the mountains and get their good tidings. Nature's peace will flow into you as sunshine flows into the trees."

Muir also had an _____ imagination and a lot of _____ that led to his becoming an inventor. However, Muir could not separate himself from _____ nature for long. He set out on a _____ of 1,000 miles.

A few years later, Muir lived and worked in Yosemite Valley, among the _____ and ancient redwood trees. He took many risks to _____ the rocks in Yosemite to gather scientific evidence that glaciers carved the valley. Muir also climbed trees in wind and rainstorms to discover how they behaved.

For more than 25 years, Muir struggled to _____ the forest and wilderness areas. He had the _____ to write persuasive articles so _____ would be interested in finding _____ to preserve wilderness for future generations to enjoy. He worked hard to _____ the damage to forests and habitats as the _____ of too much logging.

One of his many _____ was influencing President Roosevelt to create five national parks. Muir _____ forests around the _____ to help preserve them. Muir's last battle was to prevent the construction of a dam in the Hetch Hetchy Valley. The _____ lasted 10 years and in the end, his _____ won.

Story of John Muir: Answer Sheet

	Key Words	
respect	wandered	observe
treasure	active	exploring
ancient	magnificent	Risks
examine	struggle	defend
ambition	reverse	triumph
solutions	surveyed	Globe
ordeal	opponent	outcome
individuals	journey	solutions

John Muir learned to love and (**respect**) nature when he was a young boy at his family's farm in Scotland. He (**wandered**) through fields and forests to (**observe**) the birds, squirrels and other wildlife. Muir always (**treasured**) nature. "Climb the mountains and get their good tidings. Nature's peace will flow into you as sunshine flows into the trees."

Muir also had an (**active**) imagination and a lot of (**ambition**) that led to his becoming an inventor. However, Muir could not separate himself from (exploring) nature for long. He set out on a (**journey**) of 1,000 miles.

A few years later, Muir lived and worked in Yosemite Valley, among the (**magnificent**) and ancient redwood trees. He took many risks to (**examine**) the rocks in Yosemite to gather scientific evidence that glaciers carved the valley. Muir also climbed trees in wind and rainstorms to discover how they behaved.

For more than 25 years, Muir struggled to (**defend**) the forest and wilderness areas. He had the (**ability**) to write persuasive articles so individuals would be interested in finding (**solutions**) to preserve wilderness for future generations to enjoy. He worked hard to (**reverse**) the damage to forests and habitats as the (**outcome**) of too much logging.

One of his many (**triumphs**) was influencing President Roosevelt to create five national parks. Muir (**surveyed**) forests around the (**globe**) to help preserve them. Muir's last battle was to prevent the construction of a dam in the Hetch Hetchy Valley. The (**ordeal**) lasted 10 years and in the end, his (**opponents**) won.

Who Was John Muir?

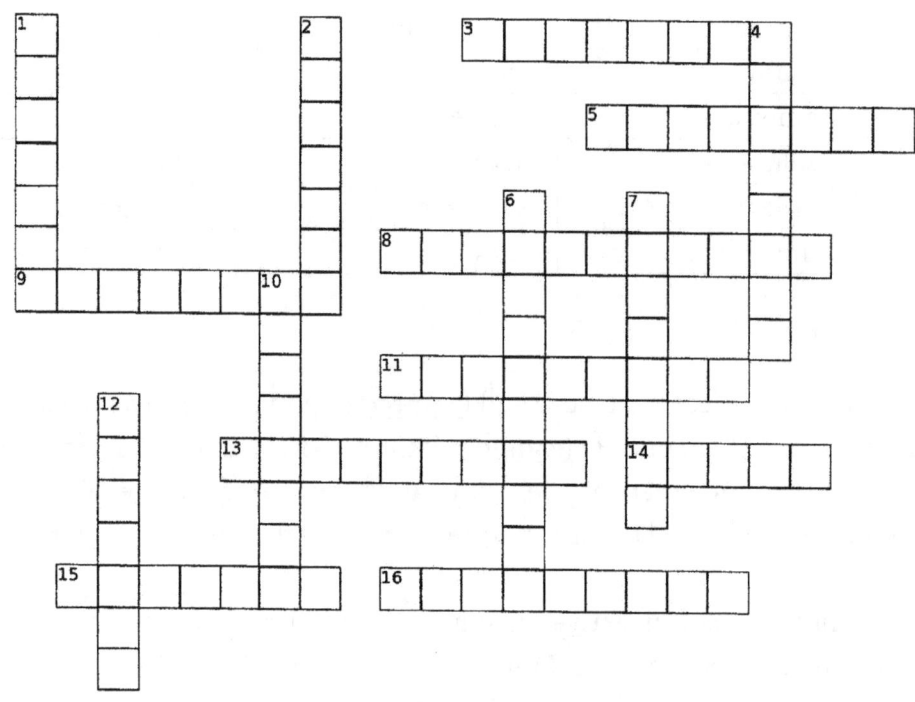

ACROSS

3 Muir had a special talent for building _____.
5 City where he lived and worked on his ranch.
8 The room where he wrote many of his books and stories.
9 One of Muir's favorite animals to watch.
11 What he was called when Muir studied glaciers and created new farming practices.
13 He called the Sierra Nevada _____, the "Range of Light."
14 This is where Muir grew fruits and vegetables to sell.
15 Muir collected and drew hundreds of these.
16 Muir and _____ Roosevelt visited Yosemite valley

DOWN

1 He spent many years surveying these to be sure the trees were healthy.
2 Where he described his adventures, recorded measurements and drew pictures of what he observed.
4 The name of his dog.
6 The special places Muir wanted to preserve.
7 Yosemite Valley was carved from these.
10 A person who studies places no one else has visited.
12 Muir believed logging the forest was a very serious _____.

Who Was John Muir?

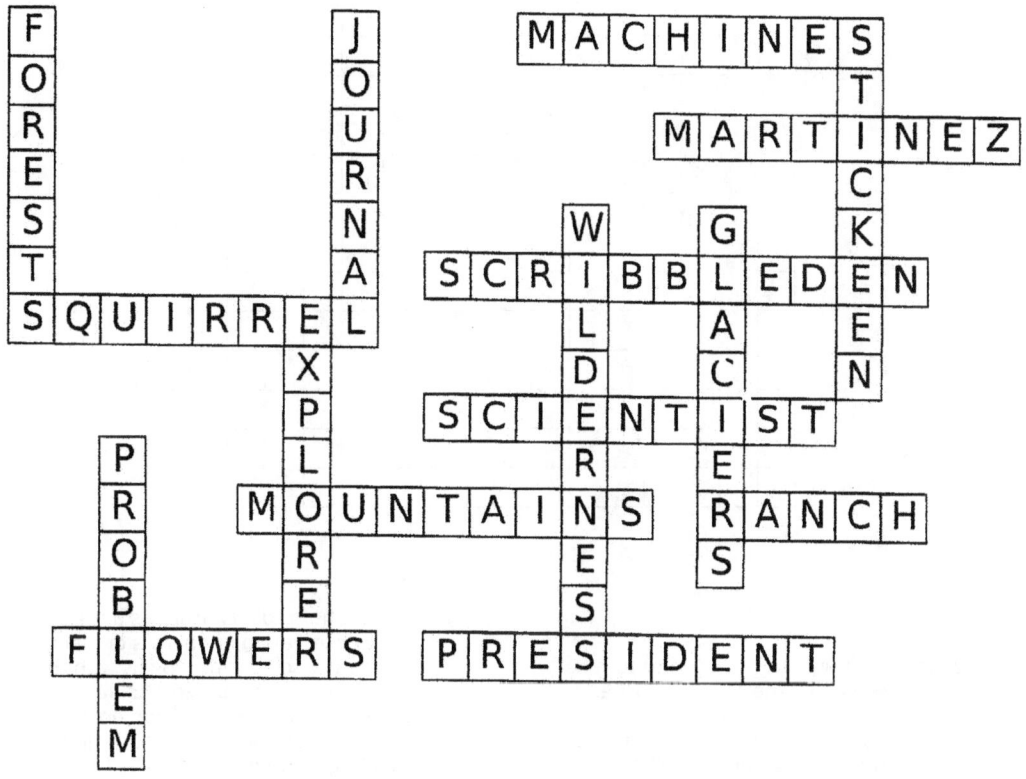

Name _____

John Muir: The Scientist

ACROSS

2. John Muir was a _____, because he studied plants, their life, structure, growth and classification.
4. The study of growing flowers, fruits and vegetables in gardens or orchards.
6. The area of science that deals with physical history and structure of the earth, rocks and rock formations is called _____.
10. A steep-sided, U-shaped valley formed by erosion when glaciers move through a river channel.
11. A large mass of ice formed when snow accumulates and stays solid without melting.
12. The physical world we live in, including things, influences and conditions.

DOWN

1. John Muir was called a _____ because he studied nature as a result of direct observation of animals or plants.
2. The area of science that deals with the origin, history, characteristics, life processes and habits of living organisms.
3. John Muir was opposed to this practice of cutting so many trees in the forest that caused long term damage to habitats and stripped the landscape.
5. John Muir founded the _____ movement that involves the careful use of resources to prevent their injury, waste, decay or loss.
7. A geographic area where plants, animals and other organisms, as well as weather and landscape, work together to create life.
8. John Muir was one of the first to apply the area of science that studies the relationships organisms have to each other and the environment.
9. Weather patterns of a specific place that occur over a period of years.

John Muir: The Scientist

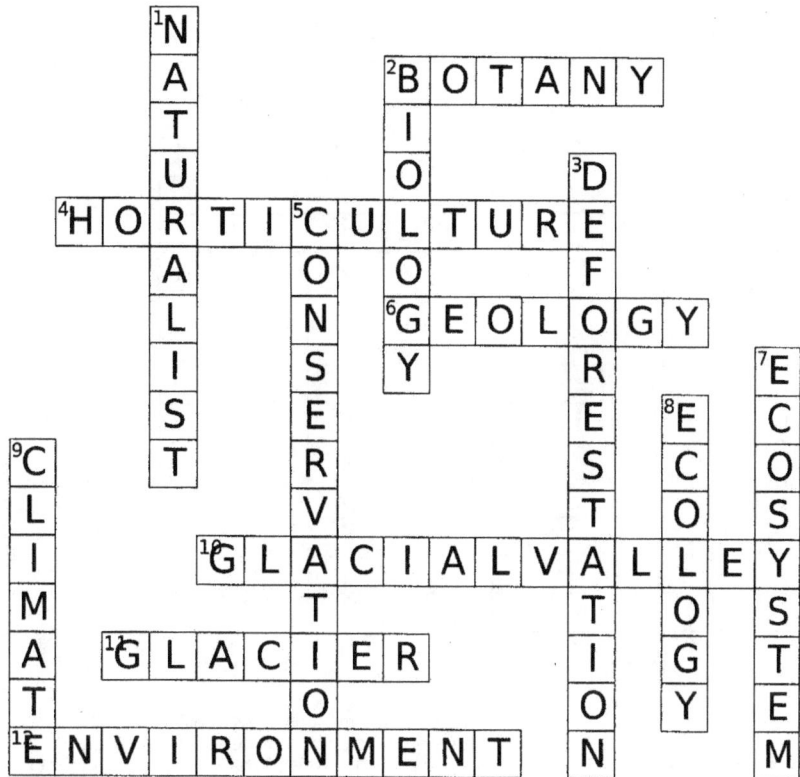

Science Crossword

ACROSS

2. A large mass of ice formed when snow accumulates and stays solid without melting.
6. A geographic area where plants, animals and other organisms, as well as weather and landscape, work together to create life.
7. The area of science that studies the relationships organisms have to each and the environment.
12. Cutting down (or clearing) large amounts of tree so quickly that the forest has no time to grow back and the land lays bare.
13. The art of science of growing flowers, fruits and vegetables in gardens or orchards.
14. A mass of loose rock, soil and earth that sits in front of a glacier as it moves down a valley and then deposited where the glacier stops.
15. The physical world we live in, including things, influences and conditions.

DOWN

1. The area of science that studies plants, their life, structure, growth and classification.
2. The area of science that deals with physical history and structure of the earth, rocks and rock formations.
3. The area of science that deals with the origin, history, physical characteristics, life processes and habits of living organisms.
4. A person who studies, or is an expert in, nature as a result of direct observation of animals or plants.
5. Name of the reservoir formed when a dam was built and flooded a valley of the same name inside of Yosemite National Park.
8. When an object can be decomposed, either by living organisms or naturally over time back into the earth.
9. A steep-sided, U-shaped valley formed by erosion when glaciers move through a river channel.
10. The careful use of resources to prevent their injury, waste, decay or loss.
11. Weather patterns of a specific place that occur over a period of years, such as cloudiness, temperature, air pressure, humidity, rainfall and winds.

Science Crossword

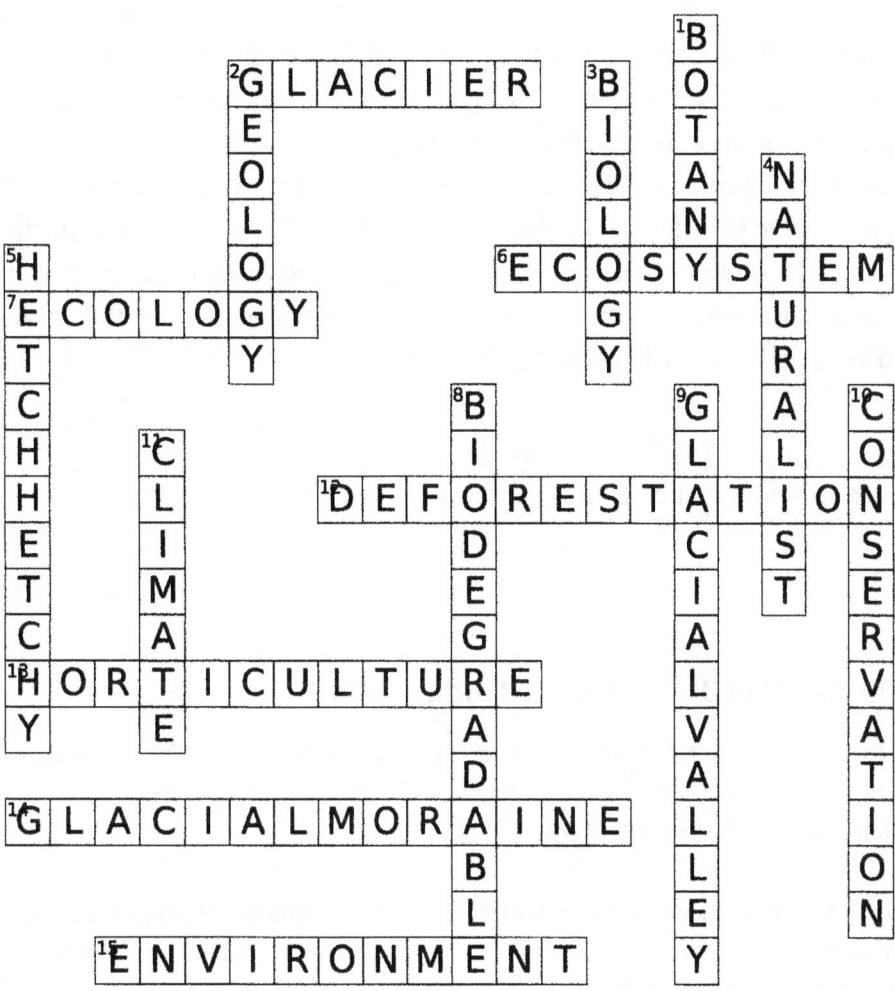

Agriculture

Background

John Strentzel and John Muir were both agricultural pioneers on their ranch. The status of agriculture when Strentzel first established his ranch was more of a "gentleman farmer," meaning he farmed more for pleasure, or to be self-sufficient, instead of to earn high profits. By the time Muir took control of the ranch, agriculture had changed and become an increasingly complex and competitive business that required sound management practices and constant attention.

Successfully managing an orchard included a number of considerations, including decisions such as the planting arrangement of fruit trees. Trees planted in uniform, straight rows made the best use of irrigation, and required less labor to prune and harvest the fruit. Labor costs were increasing and more uniform rows required fewer hours to maintain, to keep prices low when selling the fruit.

Goal

Introduce students to the business and art of agriculture.

FAST FACTS ABOUT EARLY CALIFORNIA AGRICULTURE

- By the end of the 19th century, the agriculture industry in California had already changed from mission agriculture to cattle, to sheep to wheat. The next change was introducing irrigated farmland in 1880 and tripled in use by 1920.

- By 1910, California emerged as one of the world's principal producers of grapes, citrus, and various annual fruits. The ability to ship produce in refrigerated railroad cars across country was one of the major reasons for this leadership position.

The Art of Agriculture

Background

Fruit labels were developed in response to increasing competition to sell a farmer's produce. Printing presses were able to produce colorful labels that set apart one farmer's product from another. See attached Brief History of Fruit Labels for full story.

Goal

Student art projects become an opportunity to discuss the artistic side of California's agricultural history.

Objectives

1. To design either a poster, crate label, postcard or logo that resembles the packing crates used during the early days of California agriculture.
2. To draft a promotional slog to promote the produce described in the labels the student has created.

Procedures

1. Briefly discuss the role of art in agricultural history, as it was used posters, labels, packing crates, tickets, postcards, magazine advertisements and other sources.
2. Search Google Images using the keywords, "Vintage Fruit Crate Art" to find images of authentic fruit crate artwork to show students.
3. Find original art for purchase from www.thelabelman.com or www.asliceintime.com and other online sellers.

Activity

1. Students create one or more drawings to use as labels, postcards, etc for their imaginary produce ranch or the Strentzel-Muir Ranch.
2. Students develop a slogan to be displayed with their product label as a promotional message.
3. Students share their completed drawings by identifying the food they illustrated and the rationale for the slogan they chose.

Follow-up and Evaluation

1. What skill level did the student demonstrate while drawing?
2. Does it appear that some students would benefit by tracing or coloring in a template instead of drawing freehand?

36 | *Practices in Environmental Stewardship*

3. Did students complete their drawings? Was time allocated to the activity a limiting factor?

Advance Preparation

- ✓ Writing and drawing supplies (pens, colored pencils, crayons or markers)
- ✓ Stencils for tracing shapes of produce and other shapes (optional)
- ✓ Images of fruit crate labels to use as samples

Common Core Standards

Third grade. Presentation of Knowledge and Ideas. 4
Fourth grade. Presentation of Knowledge and Ideas. 4

Brief History of Fruit Crate Labels

Fruit labels were originally created when produce was shipped by rail in wooden crates after the transcontinental railroad was completed. So many new markets opened up that farmers were competing for places to sell their fruits and vegetables. Crate labels were a way to identify the farmer's products and create the impression that one farmer's food was better than another one.

When John Muir began ranching in 1880, the country was already changing from a local market economy – where people either grew what they need themselves or purchase what they need in their own town- to a mass merchandise economy. In a mass-market economy, goods people need are produced in one place and sent across the country or the world for people to buy.

The railroad featured refrigerator cars that shipped produce on ice across country. John Muir agreed to have a train trestle built on his ranch and wanted his own train station. This was his way of being better than the competition because he traveled a very short distance to ship his fruit. There was less chance of it spoiling before his products reached the train.

People living on the East Coast loved getting California oranges during the winter. For them, it was like enjoying a ray of California sun with every bite. People across the country were hungry for California oranges. In 1893, the Southern California Fruit Exchange was born. You know the company now as Sunkist Growers.

Several other things were happening at the same time. The solder that seals metal cans together was improving. When cans did not leak, the food stayed fresh. Shipping food in cans helped meet the increasing demand for California foods. Canneries began to open in California. There were canneries in Martinez and Monterey to can seafood. There was also lot of canneries in Sacramento. The only cannery left now processes tomatoes.

There were advances in printing presses. Printers could reproduce labels with colorful pictures of apples, oranges, flowers or farm scenes.

The railroad opened up so many new markets for California goods, that competition increased. When farmers started looking for ways to identify their produce and separate themselves from other farmer's products, labels began to be important.

The fruit ranchers were the first to recognize the benefits of advertising to mass markets by using their labels. They understood that using a brand name meant a lot to consumers. When consumers saw a farmer's label, they associated the name with being a good quality product.

Today, we see brand names everywhere. We wear brand names on our shoes, our jeans and our shirts. We buy food at places with brand names, such as like McDonalds , Starbucks or Jamba Juice.

What happened to fruit labels?

Between 1885 and 1956 there were millions of labels printed. In the 1950s, farmers stopped using labels. Heavy-duty cardboard boxes featuring pre-printed pictures replaced wooden crates. The boxes were far less expensive to produce than crates. Fruit labels are now collectible items, similar to collecting stamps or bottle caps.

Student Bibliography

Audio recordings

Gilchrist, G. (2000). My life of adventures. John Muir. On CD. Dawn Publications: Nevada City, CA

Gilchrist, G. (2000). Stickeen and John Muir's other animal adventures. On CD. Dawn Publications: Nevada City, CA.

Books

Dunham, Montrew. (1998). John Muir, Young Naturalist. Aladdin Paperbacks: New York, NY.

Dunlap, Julie. (2004). John Muir and Stickeen: An icy adventure with a no good dog. Northword Press: Chanhassen, MN.

Faber, Doris. (1991). Nature and the environment: Great Lives. Charles Scribner's Sons: New York, NY.

Lasky, K. (2006). John Muir: America's first environmentalist. Candlewick Press: Cambridge, MA.

Mcully, E. A. (2003). Squirrel and John Muir. First ed. New York, NY: Farrar Straus Giroux. (suitable for younger readers)

Muir, J. (1998). John Muir and the brave little dog. Dawn Publications: Nevada City, CA.

Wadsworth, G. (1992). John Muir, Wilderness Protector. Lerner Publications: Minneapolis, MN.

Wadsworth, G. (2009). Camping with the president. Calkins Creek: Honesdale, PA.

Warrick, K. C. (2002). John Muir: Crusader for the wilderness. Enslow Publisher: Berkeley Heights, NJ.

Videos

Biography of John Muir, No. 1 "A glorious journey" 10 min. view at
    http://www.YouTube.com/watch?v=-CDzhIvugw8

Biography of John Muir No. 2 12 minutes view at
    http://www.YouTube.com/watch?v=Tpgx-LkvHGE&feature=related

John Muir and Yosemite 1:39 min. view at http://www.YouTube.com/watch?v=4M-Z12QRDwE&feature=related

Lee Stetson, Keeping the spirit of John Muir. 7:10 min. view at
    http://www.YouTube.com/watch?v=nVJ-RO8_v4c

Teacher Bibliography

The resources listed below provide additional background on John Muir, environmental issues, curriculum ideas and other learning opportunities.

Literature

Brooks, P. (1980). Speaking for nature: How literary naturalists from Henry Thoreau to Rachel Carson have shaped America.

California Foundation for Agriculture in the Classroom. (n.d.) Gardens for learning: Linking state standards to your school garden. http://www.cfaitc.org.

Castella, K., Boyl, B. (2005). Discovering nature's alphabet. Berkeley, CA: Heyday Books.

"Conservation." (2008, Apr.) 18(4). Kids Discover.

Cornell, J. (1987). Listening to nature: how to deepen your awareness of nature. Nevada City, CA: Dawn Publications.

"Ecology." (2002, Feb.) 16(2). Kids Discover.

Giono, J. (1995). The man who planted trees. White River Junction, VT: Chelsea Green Publishing Company.

Harwell, K. and Reynolds, J. (2006). Exploring a sense of place. Palo Alto, CA: Conexions.

Kaye, C. B. (2004). The complete guide to service learning. Minneapolis, MN: Free Spirit Publishing, Inc.

Leslie, C. W. (2003). Keeping a nature journal. (2nd ed.) North Adams, MA: Storey Publishing.

"Maps." (2000, October). 10(9). Kids Discover.

Muir, J. (2011). My first summer in the Sierra. New York, NY: Houghton Mifflin Harcourt Publishing Company.

Muir, J. (1988). The mountains of California. New York, NY: Dorset Press.

"Plants." (2006, May). 16(5). Kids Discover.

Ravilious, K. (2009). Power: ethical debates about resources and the environment. Smart Apple Media: Mankato, MN. (From the series Dilemmas in Modern Science from NoveList).

Righter, R.W. (2005). The battle over Hetch Hetchy. America's most controversial dam and the birth of modern environmentalism. New York, NY: Oxford University Press.

Roa, M. (1993). Environmental science activities kit. San Francisco, CA: John Wiley & Sons.

Zwinger, A.; Tallmadge, J.; Leslie, C.W.; Wessels, T. (1996). Into the Field. A Guide to Locally Focused Teaching. Great Barrington, MA: The Orion Society.

Organizations

Acorn Naturalists. www.acornnaturalists.com (online source for science and environmental education supplies)
California Forest Foundation. http://www.calforestfoundation.org/
California State Parks. Adventures in Learning. http://www.parks.ca.gov/?page_id=735
Children and Nature Network. www.childrenandnature.org
Environmental storytelling. www.environmentalstorytelling.org and nature story.org
National Park Service. Explore Nature. http://www.nature.nps.gov/learningcenters/
National Storytelling Network. www.storynet.org
National Wildlife Federation. www.nwf.org
School Garden Wizard. http://www.schoolgardenwizard.org
Sierra Club links to John Muir information, writing and educational materials.
    http://www.sierraclub.org/john_muir_exhibit/writings/default.aspx.
    http://www.sierraclub/john_muir_exhibit/educational_resources
Storytelling Association of California. http://www.storysaac.org/

Websites

Leave no Trace Center for Outdoor Ethics. Programs and principles. http://www.lnt.org/programs/principles.php

Field Study Activities

Chapter Two

- Nature Scene Investigations (NSI)
- Nature Games

Nature Scene Investigations (NSI) — Chapter Two

Background

Lessons within this theme create a base for inquiry-based learning opportunities, observation and decision-making. Each activity begins with the assessment of knowledge, followed by background information and preparation, the activity and follow up sharing and/or discussion as a tool for evaluation.

Goals

- Present a series of outdoor experiences that both inspire and engage students as they learn and apply concepts associated with the natural world in a way similar to how John Muir observed and described the world.
- Help build reading, writing, oral presentation, problem-solving and analytical skills while they practice working both independently and in collaboration with a partner or group.

Nature Scene Investigations (NSI)

- Meet a Tree
- Explore a Special Place
- Creek Study

Nature Games

- Recipe for a healthy forest
- What am I?

Time: 45-60 minutes

Meet a Tree

Goal

Students will observe and learn about the dynamic growth systems, structure and canopy of a single tree.

Objectives

1. Examine and take note of the tree bark, branches, canopy and structure to speculate on its age and evidence of disease.
2. Compare and contrast the branches, trunk and canopy of the tree to their human form and function.
3. Observe what lives in a tree and evidence of animals past or present.
4. Create a detailed illustration or description of their tree.

Time: 45 minutes

Procedures

Simplified Project: Invite students outside to draw a tree and what lives on it. Ask "Meet a Tree" study questions while they draw and encourage discussion.

Assessment

Program leaders identify knowledge students possess about systems of a tree.

Background – full project

1. Leaders introduce the idea of "meeting a tree" through careful observation. They describe the living systems of a tree (bark, sap, leaves, canopy) and ask students to consider and provide feedback on how their own living systems are the same or different than trees.
2. Students are asked to examine both the "big picture" of the tree - size, structure and shape of the tree canopy and the small details that may include bark texture, signs of wildlife, disease or injury.
3. Students are instructed to select their own tree and record their observations, based on the attached study questions, through illustrations, narrative, and/or data measurements in field journals.
4. Hand out study questions. Review the questions to assure that all students understand the questions and what they are expected to do.

Activity

1. Students may go individually or in pairs to meet their tree. Each student is responsible for recording individual observations in his or her own journal.
2. Students are called to return to the station after 30 minutes of observation based on the questions in their handout in addition to observations of their own.

Follow up and Evaluation

- Leaders engage in a brief discussion of what students observed and recorded.

Advance Preparation

- ✓ Assemble field journals

Equipment and Supplies

- ✓ Field journals or blank paper and clipboards
- ✓ Pens, pencils or colored pencils
- ✓ Measuring tapes and hand lens (optional)

Follow-up Activity

1. Interview an Urban Forester to learn more about trees and how they live
2. Plant a tree at school or in the neighborhood

References

Cornell, J. (1998). Sharing nature with children. Nevada City, CA: Dawn Publications.
Linnea, A. "Meeting a Tree," Retrieved from: PeerSpirit.com January 28, 2012.

Common Core Standards

Third grade. Text Type and Purposes. 2. Presentation of Knowledge and Ideas. 4, 6
Fourth grade. Text Type and Purposes. 2. Presentation of Knowledge and Ideas. 4, 6

Next Generation Science Standards

Biological Evolution: Unity and Diversity 3-LS4-3, LS4.C: Adaptation, Argument from Evidence, Cause and Effect

Meet a Tree Study Questions

The questions below are guides to help "meet the tree" and create an understanding of its life and habitat. Respond thoughtfully to as many questions as time permits. Do not feel pressured to answer every question.

1. How does the tree receive and transport its nutrients to survive?

2. Describe the tree bark. (use descriptive words such as density, texture, scent or color)

3. What is the usual habitat of the tree? (such as forest, wetland or grassland) How has this tree adapted to its habitat? In what way? In what habitat(s) do you suspect it would not survive?

4. Draw a leaf or needles from your tree. Describe how the leaves/needles arranged on the branch?

5. Describe the role of the leaves or needles in supporting the life of the tree.

6. Look at the ground near the tree for signs of decomposition. Describe the evidence.

7. Describe the difference between humans and trees in the way food is transported through our respective "trunks."

8. How do you think this tree reproduces itself?

9. Name three ways or living things that can harm this tree. (not including cut down by a saw)

10. Visit a second tree that has different characteristics. Compare the bark, tree canopy, leaves or needles and cones.

Explore a Special Place

Goal

To carefully examine one microhabitat to discover what life forms live and travel through a confined space and their chance of survival in that habitat.

Objectives

1. Measure and mark a confined area of 1 square yard to examine as a work team.
2. Use field study equipment (e.g. hand lens, measuring tools, tweezers) to identify, examine or analyze at least five living creatures or phenomenon situated or through a confined space.

Time: 45 minutes

Procedures

Assessment

The leader asks the students if they have studied a square yard and watched creatures moving in and out. If so, what did they see?

Background

1. Ask students to work in pairs or three to observe a small, contained space to discover what creatures live there and what other activity takes place. Encourage students to look for certain objects or animals, such as ants crawling on a flower, or a trail left by a snail, or differences in soil color or texture.
2. Special places must be somewhere that are close enough to see small objects clearly and can be marked off in some way. This does not necessarily need to be on flat ground. Students need to mark quickly because they have a limited time (30 minutes) for observations.
3. Show a sample page of Muir's journals to demonstrate his observations
4. Pass out the "Explore a Special Place" study questions and review with students to be sure they understand the questions and what they need to do.
5. Distribute student field kits containing supplies to mark their area.

Activity

1. Students select their special place to explore, measure one-foot square area. If they choose to mark the ground, they will use string small rocks or weights for the corners. If marking off the ground, they can use blue removable tape.
2. After marking the site, students describe, draw, measure, identify and examine what they see in such a way that it would be possible for other student to understand the habitat
3. Students are called to return to the station with supplies after 30 minutes of observation based on the study questions they received in addition to their observations.

Follow up and Evaluation

- Leaders engage in a brief discussion of what they observed and recorded in their journals.

Advance Preparation

- ✓ Create NSI Field Journals, use clipboard and paper or pocket folders and paper
- ✓ Secure stones as weights
- ✓ Cut string to desired lengths
- ✓ Secure rolls of blue painter's tape from hardware store

Equipment and Supplies

- ✓ Field Journals, Pens, Pencils
- ✓ Small stones, string
- ✓ Tape measure and Blue painter's tape

Extensions

- Revisit their special place in another month or two to see if the area has changed, how it is different and what new objects they can find.
- Identify a specific place at school where they can monitor activity on a weekly basis
- Engage in classroom research to identify how a specific microhabitat experiences and adapts to seasonal change

Reference

Herman, M.L., Passineau, J.F., Schimpf, A.L. Schimpf & Trener, P. (1991). Teaching kids to love the earth. Duluth, MN: Pheifer-Hamilton.

Common Core Standards

Third grade & Fourth Grade. Text Type and Purpose. 2. Presentation of Knowledge and Ideas. 4,

Next Generation Science Standards

Biological Evolution: Unity and Diversity 3-LS4-3, LS4.C: Adaptation, Argument from Evidence, Cause and Effect

Explore a Special Place Study Questions

The questions below are guides to help begin your exploration of a microhabitat. Respond thoughtfully to as many questions as time permits. Do not feel pressured to answer every question unless you have time to consider each of them carefully.

1. What life forms or evidence of life do you see in your special place?

2. Is this place where living organisms can live or is this place where they eat and go back "home?"

3. What changes can happen to this place that would change the life forms chance of survival? What adaptations would the organism need to make to survive in such an environment?

4. Do you think a living organism's size has any influence on its survival in this space? What do you think is the largest size a living organism could be?

Creek Study

Goal

Study creatures that live in a creek habitat and develop a hypothesis for how their lives could be influenced by changes in habitat.

Objectives

1. Identify creatures living in the creek.
2. Identify the water source for the creek.
3. Brainstorm ideas on what forces of nature and/or predators can change the habitat.
4. Develop a hypothesis on the impact of change to the environment.
5. Conduct investigations on a phenomena occurring at the creek.

Time: 45 minutes

Procedures

Background

1. Explain that creeks provide habitat for creatures living in and around the creek.
2. Provide two or three examples, such as weather, pollution or erosion that influence the life of the creek and the creatures that depend on it for their survival.
3. Ask students to conduct an investigation of the creek in pairs or groups of three.
4. Describe procedures for observing the creek using the hand lens, water sample cup and their own senses of hearing, sight, smell and touch.
5. Advise students that they need to decide on an experiment to conduct where they can predict what will happen, record the data and repeat the experiment two three times to verify results.
6. Pass out Creek Study Questions handout to guide students when developing and recording their experiments. Review the handout to be sure that students understand project instructions.
7. Describe methods for recording observations their NSI field journals based on the Creek Study questions. (tables, graphs, lists, narrative descriptions).

Activity

- Leader or students form pairs or groups of three and select a spot on the creek to study.
- Leader monitors activity, prompts and/or responds to questions as needed.
- Leader calls students back after 30 minutes of investigation.

Follow up and Evaluation

- Leader engages in a brief discussion of student investigations, predictions, findings and challenges.

Advance Preparation

- ✓ Assure that the creek offers areas where students can make at least three or four meaningful investigations.

Equipment and Supplies

- ✓ Thermometer
- ✓ Tape measure
- ✓ Container to collect samples (e.g. plastic bag or clear cup with lid)
- ✓ Hand lens or magnifying glasses
- ✓ Field Journal

Reference

Roa, M. (2011). The Conifer Connection: A guide for learning and teaching about coniferous forests and watersheds. California State Parks. Retrieve at www.caltrees.org.

Common Core Standards

Mathematics - MP.2 Reason abstractly and quantitatively, MP.4 Model with mathematics, MP.5 Use appropriate tools strategically

Next Generation Science Standards

Earth's Systems 3-ESS2-1, 4-ESS2.A
Analyzing and Interpreting Data, Weather and Climate, Patterns

Creek Study questions

The questions below are guides to help begin your investigation. Develop an experiment where you can predict what will happen. Repeat your experiments two three times to verify your results.

1. What do you notice about the water? What is its color and temperature? Do you see sediment in the water? If so, how much? What creatures do you see living in the water? What is the speed of the water and its level?

2. Describe ways that you can estimate the speed of the water. Think about what you can hear, feel or see in the water that gives you clues to its speed. What clues can you find that might slow the speed of the water?

3. After you have looked at the water with your eyes, pick up a hand lens or magnifying glass and check if you can see something else that you missed - something was too small to see with your eyes alone.

4. Do you think this creek is the only place where the creatures in this habitat can survive? Do you notice any signs of creature adaptation? If so, what?

5. How does the water temperature in one place compare to another place on the creek? What do you think is the reason for this difference?

6. Where does the water in the creek begin?

7. Develop a hypothesis about why or how this creek can change. What could happen to the surrounding environment if the condition of the creek changed?

8. Develop a graph or table to record findings of your investigations.

Nature Games: Recipe for a Healthy Forest

Goal

Understand the interrelationships of elements that compose a healthy forest.

Objectives

1. Illustrate at least 6 parts that make up a healthy forest.
2. Understand the importance and play the role of one element in the group's forest.
3. Recognize the problems that can occur when one or more elements are altered in a forest habitat.
4. Engage creative thinking skills when student create their dream forest.

Time: 45-60 minutes

Procedures

Assessment

1. Identify students who have visited a forest.
2. Ask one student to describe a forest and what lives there in a few words

Background

1. Instruct students to draw a picture of what they think a healthy forest looks like. Emphasize that healthy forests are an ecosystem and can include trees, animals, waterfalls, good soil, decomposers, producers and different types of weather.
2. Tell students to be specific about the names of elements. Instead of saying decomposer, say the name of the decomposer, "ants."
3. Discuss with students the need to consider their behavior when inside a forest. Healthy forests depend on the actions people take inside and outside a forest.
4. If time allows, conclude with a discussion of seven Leave No Trace principles.

Activity

Visioning a healthy forest

1. Students work alone to envision their forest for 15 minutes.
2. Leader stands by to answer questions and monitor individual progress.
3. Leader calls students back into group.
4. Ask all students to share their illustrations and say a sentence or two about them.
5. Ask students what their drawings have in common with each other.

Preparation

1. Ask individual students to choose the part of the forest in a group game.
2. Assure all the critical parts of a healthy forest are assigned to one or more students (trees, waterfall, rain, good soil, etc.)
3. **Discuss with students what is really necessary for a forest to sustain its health.** One important element of a healthy forest is its diversity. A diverse combination of life forms (as part of a complex ecosystem) is vital to the health of a forest. The interdependence of the different species and processes that take place in a forest help to maintains forest health. Wildlife depends on healthy forests for their habitat and contributes to the interwoven world that keeps a forest strong. People depend on forests for food, fiber, recreation, water quality and economic stability. Unfortunately, when the air is filled with pollution, the bad air can harm trees and make it unhealthy for everyone who lives and works in the forest.
4. Give students a sign to wear around their necks naming the role they are playing
5. Set aside students to each wear one of the following signs: **person**, **fire** and **trash**.
6. Ask students to describe their role, then take their position in the forest. Students will be moving if their role (waterfall, rain, etc) dictates they do so. Other parts might be still, such as a tree or soil, so students need to act accordingly.

Playing the game

1. Advise students that they will be in character for two minutes. Then the leader will call "freeze."
2. Wait 2 minutes, tell students to freeze, add **person**.
   - Ask students if **person** will make a difference in the forest
   - Does it matter if it is short or long? Or where it is placed?
3. Tell students to move again for 2 minutes, then freeze and add person next to trail.
   - Ask students if **person** will make a difference in the forest
   - What if person is on the trail or off the trail? What's the difference?
4. Instruct students to move again for 2 minutes and freeze. Add trash to person
   - Ask students what difference **Trash** makes in the forest
5. Instruct students to move again for 2 minutes and freeze. Remove trail and add fire.
   - Tell students that **person** has gone off the trail and made a **fire**. What difference will make in the forest?
   - What if **person** is gone and lightning caused the **fire**. What can happen in the forest?

Follow up and Evaluation

- Call on individuals to identify one thing they learned about healthy forests.
- What things change the forest so it can become unhealthy?

Preparation

✓ Make hanging nametags in advance for forest elements.

Equipment and Supplies

✓ Paper, colored pencils, markers
✓ String and card stock for hanging name tags

Reference

Cornell, J. (1998). Sharing nature with children. Nevada City, CA: Dawn Publications.

Extensions

- In classroom, use the same activity and create props, masks, antlers, etc instead of using hanging nametags.
- Build the game into a play with a narrator and audience participation.
- Create recipes for a healthy river, neighborhood, school or garden.
- Discuss and do art project focused on seven Leave No Trace Principles for Kids

    http://lnt.org/learn/7-principles

Know before your go	Choose the right path
Trash your trash	Leave what your find
Be Careful with fire	Respect wildlife
Be kind to other visitors	

    "The member-driven Leave No Trace Center for Outdoor Ethics teaches people how to enjoy the outdoors responsibly. This copyrighted information has been reprinted with permission from the Leave No Trace Center for Outdoor Ethics: www.LNT.org"

Common Core Standards

Third grade. Presentation of Knowledge and Ideas. 1, 2, 3, 4, 6
Fourth grade. Presentation of Knowledge and Ideas. 1, 2, 3, 4

Next Generation Science Standards

Biological Evolution: Unity and Diversity 3-LS4-3
    LS4.C: Adaptation, Argument from Evidence, Cause and Effect
Earth and Human Activity 3-ESS3-1, 4-ESS3-2
    Analyze and interpret data, Weather and climate, Patterns

Nature Games: What am I?

Goal

Engage students thinking about the individual characteristics of wildlife and natural resources in a playful way.

Objectives

1. Engage their imagination and spirit of fun
2. Learn about objective characteristics in playful way

Time: 45 minutes

Procedures

Preparation

1. Program leader advises group they will be playing a "What am I?" guessing game.
2. Leader divides students into two teams. Each team takes an outdoor related name such as the nature team, lion team or wild team. Leader may also assign team names.
3. Leader decides what team will guess first; then selects a student volunteer (who is a good reader) from the second team who reads the clues.
4. Leader shows the answer and the clues to the reader.
5. Student reader advises the group in play to listen carefully to hints. Each team select a single student to answer when called after all the clues have been read.

Activity

1. Student will read the clues one by one.
2. After student mystery reader has given all the clues, he or she asks for the answer from the team.
3. Encourage students to work together and assign one person to give the answer.
4. Teams switch places. Leader selects a mystery reader from the first team to read to the second team and the process repeats.

**This guessing game is a favorite for all students, played as class group or on teams. Great way to fill extra and "squirmy" time between class activities, before lunchtime or recess.

Follow up and Evaluation

- Are all students getting a turn to guess?
- Are clues too hard or too easy?

References

Cornell, J. (1998). Sharing nature with children. Nevada City, CA: Dawn Publications.

Oracle Thinkquest Educational Foundation. Retrieved from http://library.thinkquest.org/J002415 March 5, 2012. "Giant Sequoia."

Common Core Standards

Third grade. Comprehension and Collaboration. 1
Fourth grade. Comprehension and Collaboration. 1

"What am I?" Clues

Clues game one

1. I am a carnivore and love to eat creatures when they are still living.
2. My mother lays thousands of eggs in the water.
3. During the cold winter I like to stay warm, so I hibernate.
4. I use my long, sticky tongue to catch my food.
5. My family likes to jump. We can jump up to 20 times the length of our body at one time.
6. Do not believe it when someone tells you, I will give you warts.
7. What am I? **Answer game one: Frog**

Clues game two

1. I can smell and hear very well. My eyesight is not as good.
2. I have a short tail.
3. My mother and I are both great climbers. We often climb trees.
4. My diet includes small mammals, insects, grasses, fruits, nuts, berries and sometimes garbage (or a camper's food that is not stored properly)
5. My color is dark and my mother is very big! She can weigh as much as 500 pounds.
6. What am I? **Answer game two: Bear**

Clues game three

1. I can be very cold in the winter. Sometimes I freeze because I cannot wear a coat.
2. I grow when it rains.
3. Farmers like me because I nourish their crops.

4. When I am big, I can be really good at hiding things, so some people use me as a place to dump their trash. That makes me very unhappy.
5. When I am very big, I can run across an entire state. When I am very small, you may not be able to see me.
6. People and fish love to swim in me. That makes me happy.
7. What am I? **Answer game three: River**

Clues game four

1. I am the fastest growing of my kind in the world and the widest.
2. My family has survived for 2,000 or 3,000 years and some even longer.
3. My seeds are very tiny.
4. The tallest one of me is 311 feet tall.
5. We are usually called Giants.
6. You can find a lot of my family growing in Sequoia National Park.
7. What am I? **Answer: Sequoia Tree**

Clues game five

1. Most of the time I live in a forest, but I can also live in your backyard.
2. I am brown with sharp points.
3. Native Americans have used me for food and medicine.
4. Squirrels and woodpeckers like to eat me too.
5. When I am full grown, I vary in size from two inches to twenty-four inches.
6. The seeds inside of me release when there is a fire to fall to the ground and grow new trees.
7. What am I? **Answer Game five: pine cone**

Clues Game Six

1. I am covered with scales and each one is a single color.
2. My life is short. My siblings live only four days to eleven months.
3. I am one of 728 species in the United States.
4. My family is found and flies all over the world.
5. My life is filled with many changes.
6. I am small enough to fit in the palm of your hand. But you need a net to catch me.
7. I sip nectar from flowers through my tongue and pollinate plants.
8. What am I? **Answer Game six: butterfly**

Clues game seven

1. I am a mammal.
2. I am either brown or gray.
3. I live either in a tree or on the ground all over the world in forests, parks or your backyard. You may have heard me arguing. I can be very noisy.
4. My teeth never stop growing. No matter how much I chew they never wear down.
5. My family's favorite foods are nuts, seeds, insects, caterpillars, berries or bark.
6. What am I?   **Answer game seven: squirrel**

Clues game eight

1. When I am full grown, I am 3-4 feet long.
2. The desert and the mountains are my homeland.
3. You may have seen my skin that I shed along a trail you were walking.
4. When people hear me, they usually get out of my way fast!
5. I am a predator that kills my prey by poisoning them with my sharp teeth.
6. What am I?   **Answer game eight: rattlesnake**

Clues game nine

1. Some people think of me as a good luck charm
2. There are 5,000 species of me in the world
3. Poems and stories have been written about me describing me as a grouch and that my house is on fire. All of these ideas are simply not true.
4. My favorite food is aphids and other insects that eat plants. I have a very big appetite. I get so hungry that I eat 5,000 aphids during my life.
5. My color may look pretty to you, but it warns predators to "Go away. I taste terrible."
6. Unfortunately, frogs, spiders and dragonflies often eat my family.
7. I have seven black spots and a shiny red, round body.
8. What am I?   **Answer game nine: Ladybug**

Clues game ten

1. When I live in the ocean, I am a beautiful silver color.
2. I live in both fresh water and seawater. That means I am Anadromous.
3. Fish, seals and bears catch my family. Fishermen catch us in nets.

4. I weigh about 25 pounds. Some of my family has lived a long time and grown very big. I know some of us have weighed more than 100 pounds.
5. I lay my eggs in fresh water when I am between 2 to 4 years old and then I die. Most of us never make it back to our home water to spawn.
6. I have many names. I am called Chinook, King, Coho or Alaskan.
7. What am I? **Answer game ten: Salmon**

Clues game eleven

1. I am the largest rodent in North America.
2. I am a herbivore. My favorite foods are bark, twigs, roots and aquatic plants.
3. The house I build for my family underwater keeps us safe during the winter from predators.
4. People call me "nature's engineer" because of my great building skills.
5. My teeth are so strong that I can chew right through a log.
6. I change the course of rivers when I build a dam.
7. What am I? **Answer game eleven: beaver**

Clues game twelve

1. My feathers are black, red, white and yellow.
2. I can live as long as 11 years.
3. My tongue has sharp points to help me catch worms. I like to grab the side of trees and search for worms.
4. Some people think I am very noisy and do not appreciate my music.
5. You may have heard my song on a tree, a utility pole, a trashcan or your chimney.
6. People usually call my music pecking. My family calls it drumming in rhythm.
7. What am I? **Answer game twelve: woodpecker**

Classroom Projects

Chapter Three

Summary of Classroom Projects   Chapter Three

Background

The orientation and field study sections of this guide are focused on the life of John Muir with an opportunity to look at the outdoor world through his eyes - including his passion for the natural world and his many accomplishments.

This third part of the curriculum program presents opportunities for students to follow in the footsteps of John Muir by learning from his current counterparts, exploring the student's own heritage and cultural influences, and becoming involved in their own neighborhood and greater community. Students will study and/or meet the people, places, events and issues that shape their community and, in turn, the students' own lives.

These activities are, in essence, culminating projects of the John Muir study experience. Projects range from simple one-day assignments to projects that may take place over a period of several days or weeks, and based on students working in teams.

John Muir described his travels and adventures through written narratives and illustrations in a series of journals. Classroom projects involve a significant amount of writing. Teachers may choose from one to all, depending on time available and interest. Each lesson stands alone. **Recommendations for students**: Assemble all material into one 3" ring binder. This will help secure the information and give the instructor a method to evaluate student progress.

Goal

Inspire students to become more aware of and actively involved with the human and natural resources of their neighborhoods and communities - through volunteering, stewardship, speaking or storytelling. Classroom projects are listed under the primary discipline of the assignment, although a project may cross multiple disciplines.

History and Language Arts Unit

- **Who am I?** Student researches personal history through interviews with family members.
- **People of the Past**. Student compares food and toys from 1880s with contemporary children.
- **Other People, Other Places**. Student explores the stories and the heritage and culture of people from places outside the United States.
- **Our Changing Community: A Community Photo Album**. Student researches urbanization and changing face of the student's neighborhood.

Science and Conservation Unit

- **Modern Day John Muir.** Student learns about contemporary professionals working in conservation, environmental studies, advocacy, or as a writer of conservation issues who carry on the legacy of John Muir.
- **My Role as a Community Steward.** Student takes an active role in completing simple and short-term conservation projects in their community.
- **Role of agencies.** Student identifies and researches local, state and/or national agencies whose mission is to care for wildlife and public lands.
- **My Role as a Community Advocate.** Student takes an active role in addressing community-based conservation issues.
- **My Vision for the Future.** Student defines the actions he/she will take to help create a sustainable future for their community.

History and Language Arts Unit

Background

John Muir was born in Scotland and immigrated as a child to Wisconsin with his family. He met many people during his travels around the United States and the world. Muir came to understand and appreciate people of other cultures and the influences of those cultures in their homelands and as immigrants to America. Chinese immigrants worked with Muir on his ranch.

People remember John Muir for many reasons. One of those reasons is because he wrote his thoughts down in his journal. The journals later turned into books and magazine articles. He also illustrated what he saw during his travels. John Muir kept these journals because he believed that describing and illustrating where he visited was the best way to share the glory of his experiences and inspire others to care and appreciate the natural world as he did.

John Muir's life was filled with many moments to remember that shaped and changed his life. Those moments included the first time he saw wildlife in the tide pools during his youth in Scotland, when he lost his eyesight in an accident in a factory where he worked, his first glimpse of Yosemite Valley, his adventures climbing Mt. Shasta and the day he clung to the top branch of a Douglas-fir during a rainstorm.

This section is designed to familiarize the student with his or her own family's story (heritage, culture and traditions) and the story of people who are unfamiliar to the student and come to the United States from other places and other times.

Goals

- Explore the student's own history by discussing stories, descendants, traditions and culture with their own families.
- Increase student awareness of people, practices and places other than their own community or home country.

Who Am I?   Ch 3. Lesson One

Background

Most people have at least one moment or experience in their life they consider memorable or significant, even though they may not understand the influence it had on their life at the time. For students this could be receiving their first pet, going on a special vacation or visiting a theme park, celebrating a birthday, getting a new pair of shoes, something funny that happened or something sad. For the student with no memory of positive experiences, or chooses not to, suggest they imagine a special day when they meet a favorite athlete or celebrity, or visit a place of their dreams.

Goal

This lesson combines the journaling activity of John Muir with the student's own reflection of a significant experience they write about and then present it to the class.

Objectives

1. Write a detailed account of an event or experience in their life in one or more paragraphs.
2. Illustrate the event or experience or secure a photo of the experience to accompany the written account.
3. Formally present the event of experience to the class.

Procedures

1. Introduce the concept of journaling to the students if they do not already use one in class or outside of class.
2. Verify that all students have a designated 3-ring binder with interior folder pockets and single sheets of lined paper in their binder to write their lessons in.
3. Ask students to think about a memorable event or experience in their lives including the written background information as desired. Advise students their memorable moment could be yesterday or at any time of their life.

4. Students write and illustrate in their journals and include as many details as they can remember.
5. Pass out the Storytelling Decision Map as a guide to give them ideas of what to write about and how to structure a paragraph.
6. **For older students**: Ask them to write longer entries that contain more detail as well as creating additional illustrations.
7. Students turn in rough draft with illustrations and photos for comments and revisions.
8. Students return a final draft.
9. When final draft is submitted, students need to be prepared to deliver their oral presentation.
10. Students may use the photos or drawings to complement their brief presentation (about 1-2 minutes for younger students).

Follow up and Evaluation

- ✓ What challenges, if any, did the student encounter as they worked through the project?
- ✓ Did students need additional time?
- ✓ Were students able to articulate clearly orally and in writing?
- ✓ Did the students need memory trigger materials (e.g. Storytelling Decision Map) or did they think of ideas and make connections for stories on their own?
- ✓ What level of additional guidance did the students need?
- ✓ Were students genuinely interested in writing about themselves? Why or why not?

Advance Preparation

- ✓ Students bring or are supplied with bring 3-ring binders, preferably including folder pockets for storage and loose sheets of binder paper.

Common Core Standards

Third grade. Presentation of Knowledge and Ideas. 4, 6
Fourth grade. Presentation of Knowledge and Ideas. 4, 6

People of the Past — Ch 3. Lesson Two

Background

John Muir's father was a significant influence on how Muir spent his childhood and the man he would become. The values, interests, expertise and cultural traditions of parents and grandparents often play a large role in shaping a child's thoughts about his or her place in the world.

The activities in this lesson familiarize students with the lives and contributions of those who lived in John Muir's day. The first activity looks at life in the 1880s - the first decade when Muir managed the Strentzel-Muir Ranch. The second activity is a family history research project.

Goals

1. Students will learn more about themselves by learning more about their family history.
2. Students will learn the influence of family heritage on their lives.
3. Develop awareness for families of the past, including their inherent value, similarities and differences.

People of the Past - 1880s Child — Ch 3. Lesson 2A

Background

Was life in the 1880s when John Muir lived so different? In the 1880s, there were no electronic toys, computers or mobile phones. People walked, rode horses and took carriages when they left home instead of taking a car, bus or train. What toys did 1880s children play with? What foods did the children love to eat? By comparing the products of 1880 to modern times, students can see obvious similarities and differences between their lives and those of children from 100 or more years ago.

The 1880s Child Worksheet is an engaging way to introduce the concept that people of the past are valuable and their contributions continue to be meaningful in modern times.

Objectives

1. Identify all products they use today that were created during the 1880s.
2. Observe similarities between children who lived in the 1880s and contemporary students.

Procedures

1. Introduce the lesson by asking students to raise their hand in response to the statement they agree with most. Record the number of hands raised for each answer on the board.
    a. Life today is pretty much the same as 1880s America.
    b. Life today is same in some ways and not the same in other ways in 1880s America.
    c. Life today very different than in the 1880s America.
2. Pass out the 1880s Child Worksheet that lists popular foods and toys created in the 1880s.
3. Read instructions on the worksheet to students. Give them 5 minutes to complete it.
4. When students have finished, ask students to raise their hand to indicate what answers they have in common with the food of the 1880s. Record their responses on the white board as shown in the attached sample table.
    a. How many have 0-4 foods checked?
    b. How many have 5-8 foods checked?
    c. How many have more than 9-12 foods checked?
5. Discuss the similarities and differences in student responses.
6. Move on to the responses for games the students marked. Label and mark responses on the white board as shown in the attached sample table.
    a. How many have 0 – 4 toys checked?
    b. How many have 5 - 8 toys checked?
    c. How many have 9 - 12 toys checked?
7. Discuss the similarities and differences in student responses.
8. Conclude remarks about this assignment by asking students if they were surprised about the number of contributions made by people of the past. Ask "Does that change their opinion of the value of people of the past?"

Follow-up and Evaluation

- ✓ Were students engaged in the activity?
- ✓ Were students surprised at the number of items they use or eat regularly were invented by people of the past?
- ✓ Did the activity create awareness that people of the past are valuable, that they made valuable contributions?

Advance Preparation

- ✓ Prepare 1880s Child worksheets.

California Content Standard

Third grade. Reading Comprehension. 2.7
Fourth grade. Reading Comprehension. 2.7

The 1880s Child: What do you have in common?

Place an "X" next to the foods or toys you like (or have played with or eaten in the past). Add up the number of "X"s and write the number for each column on the TOTAL line at the bottom of the column.

_____	Foods, Snacks and Sweets	_____	Toys and Games
_____	Ice Cream Sundae	_____	Marbles
_____	Hamburger	_____	Chinese Checkers
_____	Tomato And Cheese Pizza	_____	Crayons
_____	Jell-O	_____	Board Games
_____	Chewing Gum	_____	Playing Cards
_____	Peanut Brittle	_____	Jumping Ropes
_____	Milk Chocolate	_____	Toy Trains
_____	Kellogg's Corn Flakes	_____	Rocking Horse
_____	Cream Cheese	_____	Wood Or Porcelain Dolls
_____	Dr Pepper	_____	Doll Houses
_____	California Oranges	_____	Spinning Tops
_____	Coca Cola	_____	Toy Boats
_____	Frozen vegetables	_____	Pick up Sticks
_____	TOTAL	_____	TOTAL

Lesson Two: People of the Past - Family History Project Ch.3 Lesson 2B

Background

Students may learn about themselves by interviewing family members: why their hair is red, they have a quick temper, a stubborn streak or the possession of a special talent. Students may substitute a neighbor or close friend if they do not have family members who are available for an interview.

Objectives

1. Interview a family member and write up a story based on the interview.
2. Use one or more images, such as photographs or illustrations, with a caption to describe a significant object or place in the life of a close family member.

Procedures

1. Discuss with students the concept that an individual's personal history, and often their future, is shaped by parents and other members of his or her family.
2. Pass out the Student Personal History Assignment Sheet.
3. Refer to the Storytelling Decision Map for ways to triggers ideas for building a simple story
4. Pass out suggested Interview Questions handout.
5. Pass out one sheet of colored construction paper, approximately 8-1/2 x 11 size to use for mounting the visual images each student collects related to their family interview.

Follow-up and Evaluation

- What is the depth and significance of information that the student presents?
- Are sentences written clearly?
- Does it appear that the student has captured the essence of the person they interviewed?
- What problems, if any, did students face when completing their assignments?
- Was the assignment meaningful to the students?
- Does it appear that they gained insight from the assignment?

Advance Preparation

✓ Reproduce the Student Assignment Sheet, Personal History Interview Question suggestions and the Storytelling Decision Map.

Common Core Standards

Third grade. Writing Standards. 2, 3
Fourth grade. Writing Standards. 2, 3

Family History Assignment Sheet   Ch 3. Lesson 2B

Select one of the assignments described below as your project.

- Interview one or more parents, grandparents, uncles or aunts and write a story about what you discovered. Use the Storytelling Decision Map if you need help to create the story.

- Draw a picture or make a copy of an old family photograph of an object of place described in the interview. Next, add a one or two sentence caption to your photo. Attach your drawing or photograph, along with your caption to the colored construction paper you received.

- After completing the interview, conduct research on what types of games and toys were used when your parents or grandparents were young. Write a summary on what you found and include photographs or drawings of the object.

Family History Interview Questions

1. When did my families arrive in the area?
2. What is the birthplace of my parents or grandparents?
3. How did my parents or grandparents meet? Were there traditions concerning courtship?
4. What work did my grandparents do? What was their mode of transportation?
5. Did they live in the city, country or farm?
6. What games did my parents or elder family members play? What did grandparents do for entertainment? Did they have leisure time?
7. Where did elder family members go to school? What was their highest level of education? What challenges did they face getting an education?
8. What work did my parents do earlier in their life? Did their work change after I was born?

Other People, Other Places   Ch 3. Lesson Three

Background

Like John Muir, people have come to live in America from other places around the world. When families come, they bring their culture and family traditions with them. America has become what is called, a "melting pot" of many nations and cultures all living together. People from other lands have different stories to share than the ones we know in America. So people bring their stories here too.

Goal

Experience the values, traditions and stories of other cultures beyond what is practiced in the student's own home.

Assignment One Objectives: Storytelling Project

1. Understand there are different types of stories that are both real and imaginary (e.g. folktale, tall tale, legend, news report).
2. Tell a story in class that represents life in another country.
3. Locate the origin of the story on a map, if possible.

Procedures

1. Introduce the storytelling project by discussing with students that cultures outside the United State have different ways of traditions and different stories.
2. Explain the different genres of stories. Ask or provide a different example of each type.
3. Read a folktale to class and describe what characteristics make that story a folktale instead of another type of story.
4. Distribute Folklore, Legends, Myths and more handout.
5. Students may go to the library and find a folktale they want to tell. Direct students that they can also visit the www.storyarts.org/library website and select a folktale from the list of stories. Click on the story they want to learn and print it out.
6. Ask students to read the story aloud with their parents or older sibling several times so they learn the basic structure and content of story and practice telling it aloud. They do not need to memorize the story word for word.
7. Set aside time in class to hear their stories and point out on a map where their story originally came from.

Follow-up and Evaluation

- Were stories delivered with clarity and enthusiasm?
- Did students learn a story to tell?

- Were students interested in the assignment?
- What challenges did they encounter?
- Did students struggle to identify and/or learn a story they enjoyed sharing?
- What did students demonstrate that they learned from the story?
- Were the stories meaningful to students?

Advance Preparation

- ✓ Locate a book of folktales to read
- ✓ Review the StoryArts.org website to identify the scope of the stories within this site.

References

Twenty-Two Splendid Tales to Tell from Around the World, Vol. 1 or Vol. 2 by Pleasant DeSpain is filled with stories and illustrations for teachers to tell.
Story Arts is located at http://www.storyarts.org/library.

Assignment Two Objectives: Interview Project

1. Research the traditions of a person from another country using primary sources.
2. Write a short biographic profile about the person interviewed.
3. Identify the country, state or city of origin of the student, the student's parents and, if possible, grandparents and mark those sites of origin on a world map.

Procedures

1. Introduce the interview project by stating the purpose is to talk to a person from another country to learn about their culture first hand.
2. Refer students to assignment sheet for project guidelines and interview questions.
3. Ask student to select a person of any age born in a country outside the United States. The person could be a classmate, a parent or other family member, a neighbor or family friend.
4. Ask the student whom they have chosen to interview. For students who need assistance in locating a person, suggest they contact a local senior center or a church in their neighborhood that represents another culture for assistance.
5. Remind students they need to conduct their interview outside of class time.
6. **For younger students**: write up of the interview can be in a Question and Answer format with one or two sentence answers.
7. **For older students**: write the interview as a one or two page biographical profile. They may supplement interview with secondary sources, including library and Internet research.
8. Post a flat map of the world on the wall. Use different colored pushpins to indicate the place of origin for the student, parents and grandparents.
9. Ask students to find the interviewee's country of origin on a map.
10. Give students time in class to present summaries of their interviews.

Follow up and Evaluation

- Does it appear that students enjoyed and learned something new and meaningful from their research?
- What challenges did they encounter?
- Did students demonstrate evidence they acquired or developed new skills?

Advance Preparation

✓ Secure source for supplies if not already in the classroom

Equipment and Supplies

✓ Hanging map of the world
✓ Three colors of push pins (at least three times the number of students in class, to account for students who use three pins to indicate country of origin)

California Content Standards

Third grade. Writing Strategies. 1.1, 1.3, 1.4, 2.1. Written & Oral Language Conventions. 1.1-1.7 Continuity and Change. 3.3.
Fourth grade. Writing Strategies. 1.1, 1.2, 1.3, 1.10, 2.1

Common Core Standards

Third grade. Writing Standards. 2, 3
Fourth grade. Writing Standards. 2, 3

Extensions

1. "Where am I," a geography lesson to map California that includes drawing in the Sierra Nevada, mountain passes, valleys, coastline and lakes. Maps can mark the destinations and the routes of travel where the student and John Muir have visited.
2. Class family history recipe book or community history recipe book. Compose the stories that go along with the food and culture.

Through the Eyes of John Muir | 75

STORYTELLING DECISION MAP

Setting
- forest
- river
- city
- pirate ship
- bakery
- zoo
- Alaska
- train or train station
- spaceship

STORY

Character + helpers or guides
- stuffed bear named Casper
- alien
- high school student
- 8 year old boy named Max
- rabbit
- coyote
- mom or dad
- family pet
- magic key
- talking cat
- talking tree
- John Muir
- Stickeen

Challenge or Issue
- earthquake, flood or fire
- polluted river / no clean water
- parent dies or parents divorce
- school newspaper closing
- friend moves away
- mother sends child into the forest with no food

FOLKLORE, LEGENDS, MYTHS AND MORE

Biography

The written story of a person's life. An autobiography is a story the author writes about him or herself.

Chronicle

A true historical record or register of facts or events arranged in the order the event happened.

Drama

A story that involves serious conflicts or challenges. Actors seen in movies or plays can appear in a dramatic production.

Fable

A fictitious story meant to teach a lesson. The characters are usually animals that talk about a situation they are facing. Aesop is the most well known author of fables.

Fairy tale

Stories of legendary deeds, princes, dragons, giants and other creatures. Cinderella, Rapunzel and Beauty and the Beast are all fairy tales.

Folktale

Includes different variations on stories and beliefs that are passed down through generations. This includes Native American stories, ghost stories, animal stories and campfire stories. Stories may be true or untrue. The author is usually unknown because the stories were passed to others as oral stories.

History

Historical stories include many different approaches that record historical events, people or a period of time. Stories can begin with actual observations or use information gathered from research to describe what happened. Usually the stories are presented in chronological order. Historical stories also record, analyze, look for relationships between and explanations of past events.

Legend

A story told verbally or in writing by many generations of people. The story cannot be proved as fact, although the story is considered to have started from facts. Often the story gets "bigger" as it is told over the years. So the listeners lose the ability to know what is real and what is imagined.

Myth

Historical stories that took place in ancient Greece or Rome that describe gods, goddesses, heroes and champions. Usually myths focus on the creation of something in the natural world, or ideals and standards of society. Why does the camel have a hump is a creation myth.

News

A general term used to describe information found in newspaper or magazine articles, TV or radio news broadcasts, or on the Internet. The story may contain background information to make the news article longer than the "news" itself. Some news contains information that is a person's opinion instead of facts.

Parable

A short, simple story, intended to express a religious lesson. Many parables are found in the Bible.

Report

This is used when an individual is describing or giving information about something they have seen or done and continues to do so over a period of time. If a person needs to be rescued and the process is a long one, then there may be several reports to describe the progress of the rescue and the person's condition.

Tall Tale

A teller tells a story that intentionally stretches the facts about real events or people to entertain the audience or the reader. These stories often focus on the achievements of a hero or even a common person who does extraordinary things. Paul Bunyan and Pecos Bill are both tall tales.

References

Denning, S. What are the main types of stories and narratives? Retrieved on 5-21-12 from http://www.stevedenning.com/Business-Narrative/types-of-story.aspx

Folklore definitions: Myths, Legends, Fables, and more. Retrieved on 5-21-12 from http://americanfolklore.blogspot.com/2005/05/folklore-definitions-myths-legends.html
http://www.americanfolklore.net Retrieved 5-21-12

Our Changing Community - Photo Album Project   Ch. 3 Lesson Four

Background

The Alhambra Valley and the City of Martinez were once a vibrant agricultural center dotted with barns, windmills and dirt roads. The original Strentzel-Muir ranch encompassed 2,300 acres in the Alhambra Valley, located in north central Contra Costa County and just outside the city limits of Martinez.

Today, urban residential and commercial development, highway and railroad corridors encircle much of the park and the remaining 326-acre ranch where John Strentzel (Muir's father-in-law) and John Muir lived and worked.

Goal

Understand the natural process of how communities change over time.

Objectives

1. Identify 5 changes in the Alhambra Valley that have happened since 1880.
2. Read and understand written historical information about the development of Alhambra Valley and the City of Martinez as an example of urbanization.
3. Participate in class discussion to demonstrate an understanding of the reasons why and how the Alhambra Valley became urbanized.
4. Relate the changes in the Alhambra Valley to the development of the student's own community.

Procedures

Reading and Discussion

Read and discuss The Little House by Virginia Lee Burton (or other picture book that illustrates urban development) as lead in to explaining the development of the City of Martinez and the Alhambra Valley.

1. Review the background of Alhambra Valley and read John Strentzel's quotation.
2. Discuss and compare the story of The Little House and the story Alhambra Valley's growth.
3. Ask students if they have ever wondered what was on the land before their house or apartment was built.
4. After hearing the story, what do they think was there before?
5. Ask students if the story matches with what they know from what they have experienced where they live or what their parents or grandparents have told them.

6. Do they know any community or family history in the area they want to share? (for example, their grandparents opened the first bakery 60 years ago.)

Community Photo Album Project

1. Pass out Community Photo Album Assignment Sheet.
2. Review the instructions.
3. Pass out 5 x 7 index cards to students (1-5 at your discretion)
4. When students return cards, display in the classroom as a community photo album with a unifying title.
5. Take class time to view and address what students noticed, found, questions and concerns.

Follow-up and Evaluation

- Were there student concerns about completing the project? Were there students who were unable to access a camera or mobile phone with camera?
- Were students able to travel safely through their neighborhood?
- What was the scope of the community featured by the neighborhood photos and captions?
- Did the photos demonstrate a balanced view of the complete neighborhood?

Advance Preparation

- Duplicate Community Photo Album Project assignment sheets
- Review the Read Aloud story to fit the attention span of your students and/or time available to read in your class

Supplies

5x7 index cards

Common Core Standards

Third grade.   Text Type and Purposes. 2. Presentation of Knowledge and Ideas. 4, 6
              English Language Arts (ELA) /Literacy   RL3.1, RL3.2, SL3.4
Fourth grade. Text Type and Purposes. 2. Presentation of Knowledge and Ideas. 4, 6
              English Language Arts (ELA) /Literacy   RL3.1, RL3.2, SL3.4

References

Martinez Historical Society
Killion and Davison. Cultural Landscape Report of the John Muir National Historic Site. (2005).
City of Martinez

Community Photo Album Project Assignment Sheet

This project involves taking a close look at your neighborhood. You will be looking for places that have changed since 1880 when John Muir came to the Alhambra Valley. Since nearly everything has changed, this will be an easy assignment!

Please take photos of places, houses and buildings, parks and open space or people that mean something to you. On a note pad that you or your parent brings along, write a caption that includes the address, name of the place or person, or a one-sentence description of the place you photograph. You will need this for this information for the project. So take photos only of what you can describe.

Attach one photo to each index card you have received. Submit as many as five photographs. Write a caption in large letters at the bottom of the index card to label the photo. Choose your best **one or two photos.** In case all the photos cannot be displayed in class, your favorite one or two photos will be posted.

Please see the teacher if you have questions or concerns about completing this project. You may use a digital camera or mobile phone to take photos, or you may draw a picture that fits on your index card.

Write Photo Ideas Here

_____

_____

_____

_____

_____

The Changing Alhambra Valley

The Alhambra Valley and the City of Martinez were once a vibrant agricultural center dotted with barns, windmills and dirt roads. The original Strentzel-Muir ranch encompassed 2,300 acres in the Alhambra Valley, located within an unincorporated area of north central Contra Costa County and just outside and south of the City of Martinez.

Today, urban residential and commercial development, highway and railroad corridors encircle much of the park and the remaining 326-acre ranch where John Strentzel (Muir's father-in-law) and John Muir lived and worked.

When John Strentzel first saw the land that he bought to establish as a his ranch, he said, "Here was a lovely fertile valley, protected by high hills, from the cold winds and fogs of San Francisco, a stream of living water flowing through it, the hills and valleys partially covered with magnificent laurel, live-oak and white-oak trees…I knew at once that the valley was well adapted to fruit growing and thought, 'here I can realize my long cherished dream of a home surrounded by orange groves, and all kinds of fruits and flowers…I immediately purchased 20 acres of the richest valley land, two and half miles from town, paying $50 per acre, and at once removed my family to the new home, they arriving on the 4th of April, 1853."

Managing the Strentzel-Muir Ranch

Managing such a large ranch was a huge responsibility for Muir. He was much more interested in spending the time he could spare between harvests to travel and work on conversation issues. He began selling and leasing parts the ranch to finance his conservation work. Over time the land was sold and subdivided, essentially "paving" the way to the urban development.

Muir was also a shrewd businessman. He gave a right of way to the Atchison, Topeka and Santa Fe in 1906 to build a railroad trestle and pass through his property. The Muir Station gave him easy access to shipping his produce. This decision "paved the way" for urban development.

Fishing industry

During the last half of the 1800s, the Carquinez Strait supported an exceptionally productive fishing industry. Thousands of pounds of salmon were shipped to destinations around the world from two canneries operating in Martinez. That industry thrived until 1957 when the San Francisco Bay was closed to commercial fishing.

Alhambra Water

Other new businesses started in Martinez and the Alhambra Valley. Loron Lassell owned a 300-acre ranch in the valley and located a fresh water springs on his ranch. In 1902, he began bottling his water and sold it under the name of Alhambra Water to San Francisco, Oakland and Contra

Costa towns. The springs were abandoned in 1954 when the company was sold to Foremost-McKesson.

Oil refineries

A major contributing factor to urban development came at the turn of the century. The desirable characteristics of deep water and rail connections attracted the attention of oil refineries. In 1895, Union Oil Company purchased land near Martinez. Other oil refineries and chemical plants followed through 1916 with the establishment of the Royal Dutch Shell. Shell Oil Company built homes for its managers and continued to expand its plant.

Changing housing and employment trends

The arrival of the oil industry sparked a real estate boom. As early as 1912, street maps of Martinez showed residential development moving in the direction of the Strentzel-Muir Ranch. Martinez developed an economic base that departed from its agricultural roots and became recognized for employment in heavy industries: oil refineries and chemical plants.

Agriculture in the Alhambra Valley

Alhambra Valley agriculture remained the dominant characteristic of the area through 1914. By the early 1920s, the lower Alhambra Valley orchards and ranches were slowly changing into paved roads and residential subdivisions. During the Great Depression, land values plummeted and farms were devastated. California farms had fared better than most, since the state was on its way to becoming a national leader in agricultural production.

New roads serve increasing population

As area population increased, more roads were built through the valley. By 1939 a two-lane road, called the Arnold Industrial State Highway was located adjacent to the Atchison, Topeka train trestle. The road split the Muir ranch's hillside pear orchards on the lower slope of Mt. Wanda. The new highway opened up the Upper Alhambra Valley for development.

World War II and forward

By World War II, California had emerged as a leader in citrus fruits, almonds and walnuts. Much of the Alhambra Valley was still used for farming. Post-war building in the 1950s introduced new residential housing subdivisions. Development continued to spread through the Alhambra Valley where farms had once filled the landscape. By the 1960s, the farms of Alhambra Valley had vanished. The population of Martinez had grown from 875 in 1880 to an estimated 36,000 in 2011.

Science, Conservation and Advocacy Chapter Three

Background

John Muir is considered America's first environmentalist. He was an accomplished inventor and engineer, studied botany, geology and biology, and several other specialties. With such a multi-faceted knowledge base, his work encompassed many contemporary professional disciplines. It is likely that naturalists, biologists, botanists, geologists and writers around the world have been inspired by the principles and actions of John Muir.

This section is designed to engage the student in local conservation activities, encourage them to take responsibility for their own actions, and create a better understanding of what advocacy, conservation, environmental and wildlife protection is about, and the role of people and agencies that are concerned about wildlife and wild places.

Goals

- Engage students in local conservation and advocacy activities so they will understand how to take thoughtful actions to care for the environment.
- Introduce students to a variety of conservation professionals and scientists that continue to build on the legacy of John Muir.
- Create student understanding of the role of various public or private agencies that care for wildlife and wild places.

Meet a Modern Day John Muir Ch. 3 Lesson Five

Goal

Students discover what role a modern day John Muir plays in caring for the environment and sharing that joy, appreciation and responsibility with others.

Objectives

1. Research and write a letter to a "modern day John Muir" of their choice.
2. Listen and ask questions of a "modern day John Muir" speaker in class.
3. Apply research and creative thinking skills to design a poster describing the work of one "modern day John Muir" professional.
4. Understand the role of several different professionals who engage in work similar to John Muir.

Procedures

1. Explain to class they will be studying a "modern day John Muir" so they can learn more about what related professionals do today. Assignment will include research through school or public library and/or the Internet, writing a letter and creating a poster.
2. Select one or more individuals listed on the assignment sheet to speak to your class about their careers and relationship to one of the roles of John Muir.
3. Inquire if there are parents of students in class working in related fields and available to speak.
4. Write a list of professionals on the board for students to select.
5. When two or more students select the same profession, those students will work on a team poster. All other students will create an individual poster.
6. Ask students to write a letter requesting information from their chosen professional. Encourage students to create their own questions – here are examples
   - What inspired the person to choose their profession?
   - Describe their job responsibilities?
   - What is the location for most of their work? Do they work in an office, in the field or in a laboratory? Do they travel long distances for their work or is it mostly local?
   - Name the special skills, interests and education needed to work in their job.
   - Do they know about John Muir? Was he ever an inspiration for them?
7. Set aside class time to write letters.
8. Ask students to bring a self-addressed and stamped envelope to send with letter for a reply.
9. Use completed posters as an opportunity to host a "career open house" for your school

Extensions

Use artifacts, lab experiments, plant identification displays or other hands-on learning experiences to complement the poster exhibit.

Follow-up and Evaluation

- ✓ Did students complete the letters and posters?
- ✓ Did the students exercise creativity and good judgment to design their poster?
- ✓ Does it appear that the students could understand and synthesize facts to present a clear, concise and informative poster?

Common Core Standards

Third grade. Conventions of Standard English. 1, 2, 3
Fourth grade. Conventions of Standard English. 1, 2, 3

Modern Day John Muir Poster Project Assignment Sheet

List of Professions

Activist (volunteer leader of local environmental organization)
*Arborist
*Biologist
*Botanist
*Ecologist
*Environmental educator
*Field researcher
*Geologist
*Horticulturalist
Lobbyist
*Master Gardener
Naturalist/Park Ranger
Rancher / Farmer
Sierra Club staff member
*Urban Forester

*For assistance locating these professionals, contact local colleges, universities, city hall or county government office for possible referrals.

Create a poster that describes the career of one professional who does work similar to that of what John Muir did.

- Research the selected modern day John Muir professional to create a poster that describes his or her career, education and skills.
- Title the poster with the name of the profession you have researched, such as "What does a Botanist do?"
- Use both primary and secondary sources, such as the Internet, parents, school books, materials found in the public library or the workplaces where the professional holds his job.
- Use descriptive words, stories, illustrations, photographs or other materials, to create your poster. Be creative and add three-dimensional items to your poster, as available.
- Posters can also compare the profession today to that of John Muir's day. For example, "Careers in Botany: 100 years ago and today."

My Role as a Community Steward — Ch 3. Lesson Six

Background

Stewardship can take many forms. This lesson continues to reinforce the concepts that a healthy community depends on 1) personal responsibility and 2) maintaining healthy forests and healthy communities for everyone is a long-term commitment.

Goal

Create awareness of the benefits of trees as contributing one part of maintaining a healthy community.

Objectives

1. Identify the role of a community steward.
2. Plant an acorn seed or participate in a tree planting in the student's community
3. List three or more benefits of trees.

Procedures

The value of a tree

1. Introduce the lesson by writing the following quote from John Muir on the board: "When a man plants a tree, he plants himself. Every root is an anchor, over which he rests with grateful interest, and becomes sufficiently calm to feel the joy of living."
2. Ask the class to describe how the quote tells what John Muir thinks about the value of trees.
3. What do the students think is valuable about trees?
4. Refer to Background on Benefits of Trees to engage in a class discussion.
5. Plan to show the movie, It's Arbor Day, Charlie Brown. This video describes the benefits of trees as Charlie Brown and other Peanuts characters prepare to celebrate Arbor Day by planting trees.
6. Discuss the movie's main points about trees. Ask students if they can say what common messages they heard in the movie and how those messages related to Muir's quote.

Invite a Speaker

Contact either the city where the school is located and ask for a referral to an urban forester or arborist to speak to your class about benefits of trees. Cities or recreation and park districts often hire these professionals to care for their trees. The speaker may also lead a tree or seed planting activity.

Sample Benefits of Trees

- Improves our air, reduces air pollution by absorbing pollutants.
- Provides shade so houses and buildings use less energy to cool.
- Improves home values because the neighborhood is more attractive.
- Controls runoff from storms because trees absorb water into leaves, bark, branches and soil. Excerpted from Sacramento Tree Foundation Leading Education and Awareness in Urban Forestry (LEAF) Program Leader's Guide, Ch. 3 (2009).

Extensions

- Contact Tree Musketeers at http://treemusketeers.org to learn how the class can become involved in a community-wide effort to plant trees.
- Walk through the neighborhood with an urban forester or arborist and conduct a "tree count" that includes a tree identification activity, and an inspection of trees to look for signs of good health and diseased trees.
- Invite speakers from various professions to speak to the class or hold a special event in combination with Earth Day. Select from the following list of professions and ask each one to address the benefits of trees in their work.

Arborist or urban forester
Homebuilder or developer
Park ranger

County tax collector
Architect / Landscape architect
Artist / Writer

Follow-up and Evaluation

- ✓ Did the It's Arbor Day, Charlie Brown movie help to reinforce the value of trees?
- ✓ Were the students able to integrate concepts from the movie with information from the speaker?
- ✓ Can students identify and understand the meaning of three main benefits of trees?
- ✓ Did students ask thoughtful questions of the speaker?
- ✓ Did students take an active part in planting trees or seeds?

Advance Preparation

Contact either the city or local tree organization to find an urban forester or arborist.

Common Core Standards

Third grade. Presentation of Knowledge and Ideas. 4, 6
Fourth grade. Presentation of Knowledge and Ideas. 4, 6

Agency Roles to Protect Wild lands & Wildlife Ch.3 Lesson Seven

Background

The federal government owns nearly half of all the land in California. Many different city, county, state and federal government agencies along with private land trusts and associations also own property in California and throughout the United States. All water in the US is owned by government agencies.

This section gives students an opportunity to find out who cares for the wild lands and the wildlife in California and the rest of the nation. John Muir is considered the father of the National Park Service. The National Park Service is the federal government agency and California State Parks is the state agency that protects wild lands, wildlife, cultural artifacts and provides visitor services.

Goal

Create awareness that public and private agencies exist to care for wild lands and wildlife and understand the agency's role in doing so.

Objectives

1. Conduct web-based and library research on one government agency that cares for wild lands and wildlife.
2. Identify the mission of the agency.
3. Identify at least 3 specific special features and/or wildlife the agency protects.
4. Create a poster using both text and images to describe the agency's role.
5. Present findings during a class presentation.

Procedures

Class discussion

1. Introduce topic of who cares for wild lands and wildlife.
2. Write the following quote by John Muir on the board: "Any fool can destroy trees. They cannot run away; and if they could they would still be destroyed — chased and hunted down as long as a dollar could be got out of their bark hides, branching horns, or magnificent hole backbones."
3. Discuss with the class what the quote means to them and why they think this quote could be important to organizations who want to protect wild lands and wildlife.

4. Write on the board the names of government agencies involved in the effort to keep Muir's vision alive. Refer to the following list for agencies concerned with wild lands and wildlife.

Student project

1. Assign students to work in pairs and choose one agency from the list of agencies.
2. Students must identify a single park site or project to research, although the mission may be described from the agency itself.
3. Set aside class time for partners to work together.
4. Students will research and create a poster that combines writing completed paragraphs with two or more images -may be copied from Internet sites or published materials from magazines, agency brochures or other information.
5. Set aside class time for students to share presentations.

Examples of government agencies to research

- ✓ Bureau of Land Management
- ✓ Bureau of Reclamation
- ✓ California Department of Fish and Wildlife (or related agencies in other states)
- ✓ California Department of State Parks and Recreation
- ✓ Individual cities and counties
- ✓ National Park Service
- ✓ US Department of Fish and Wildlife
- ✓ US Forest Service
- ✓ US Army Corps of Engineers

Follow-up and Evaluation

- ✓ Did the poster accurately reflect what the agency or site does?
- ✓ What problems did the students encounter during their research?

Advance preparation

- ✓ Secure poster board and art supplies

Supplies

- ✓ 18 x 20 white poster board

Common Core Standards

Third grade. Writing Standards. 2, 3
Fourth grade. Writing Standards. 2, 3

Agency Role Assignment Sheet

This project involves working with a partner. One partner will conduct the research and the other partner will prepare the presentation. The partners will work together to write up the findings and create your poster.

Examples:

    City government is **Martinez or San Francisco**

    County government is **Contra Costa or Alameda**

    Regional government is **East Bay Regional Park District**

    State government is **California State Parks**

    Federal government is **National Park Service**

1. Identify and research one of the branches of government that is charged with caring for wild lands and wildlife.

2. Using either the Internet or the library resources learn about the government agency you have selected. Please select one park site to research. You can use the mission statement from the agency that manages the park site.

3. You will create a poster about 18 x20 that includes descriptive paragraphs about the history of the site, special features and the wildlife that lives there. Your poster also needs to include two or more pictures of the wildlife and the site. You may use original photos, if you have visited the site, or pictures from brochures, magazines or the web page for that park site. Examples of special features include: volcanoes, scenic trails, redwood trees, lakes, rivers or waterfalls.

My Role as a Community Advocate   Ch 3. Lesson Eight

Background

John Muir proved by his actions that one person can make a difference. He was instrumental in saving millions of acres of wild lands across America. Muir is known as the "father of our national parks" in honor of his tremendous work to save millions of acres of forested land from destruction by commercial interests.

He convinced President Theodore Roosevelt to camp with him in Yosemite to share the grandeur of the park to persuade him that this wild place, and eventually many others, in America needed to be preserved for future generations to enjoy.

Muir wrote hundreds of magazine articles, and many books and illustrated journals as his way of sharing his outdoor experiences with the world. He hoped that by sharing the wonders of the natural world, people would appreciate and work to preserve it for years to come.

Goal

Engage students in exploring contemporary environmental and conservation issues. By becoming involved in advocating for or against an issue, students will come to understand the influence of their individual and collective voices.

Objectives

1. Identify one contemporary environmental or conservation issue facing their community where an individual can take action to help minimize or resolve the problem.
2. Write a one-page essay describing the problem and possible solution(s) to present to class. (one or two paragraphs for younger students)
3. Identify and take action one individual to assist in solving the problem or issue.

Procedures

1. Introduce the concept of being a community advocate. Explain that being an advocate means sharing information about an issue or project the student cares about or joins others to take action to help solve a problem.
2. Write the following quotation from John Muir on the board: "The battle for conservation must go on endlessly. It is part of the universal warfare between right and wrong."

3. Review the attached assignment sheet with students asking them to research, write, comment on a solution and take action wherever possible.
4. Encourage them to take action on their problem or issue (write a letter, attend a rally, participate in habitat restoration, etc.)
5. Allow time for students to present their findings during class

Follow-up and Evaluation

- Did students understand the concept of "issue" and "problem" related to the environment?
- Was there parental support for researching issues?
- Did students clearly articulate the issue?
- Was there ample opportunity for students to become involved?
- Did this activity create an awareness of the value of one individual's contribution?

Advance Preparation

✓ Prepare handout assignment sheets

Common Core Standards

Third grade. Writing Standards. 2, 3
Fourth grade. Writing Standards. 2, 3

Extensions

1. Form student teams to create a simple two or four-page newspaper or newsletter per team. The publication would describe the issues or problems the students researched and the actions they took to solve them. To share the issues the newspaper could be circulated throughout the school or posted on the school website.
2. Debate of opposing sides of an issue or problem.
3. A conservation issue fair in combination with Earth Day activities to display posters, essays or slide shows that explain benefits and consequences of "hot" issues involving the school, community, region or state.
4. A blog or website that keeps track of issues many of the students at the school are interested in following.
5. Participate fully in community engagement and outreach activities, such as coalition building or building partnerships with other student groups or grades to jointly address community issues.

Advocacy Assignment Sheet

This project is designed to help you identify and understand current environmental or conservation issues or problems. You may need to ask your parents for assistance identifying an issue to research and take you to the library to find books and other reference materials unavailable on the Internet.

1. Ask your parent to help you identify an important environmental or conservation issue from the newspaper, TV news or local magazine. You may also attend a city council meeting, county board of supervisors or park commission to get ideas. One idea could be writing a letter to the city council about playground equipment at a park that needs to be repaired.

2. After you have selected an issue to study, discuss it with your teacher before starting the research. Be sure the issue or problem you are interested in provides an opportunity for you to take action.

3. Go the library to read newspaper articles, magazines or search the Internet to find out more about the issue or problem. What agency or agencies are most likely to take care of the problem that you can contact for information and assistance to resolve the issue?

4. When finished with your research, write an essay describing what you found and what you think needs to be done about it.

5. If possible, take the action you have identified you can do to resolve the problem.

6. Present your findings to classmates.

My Vision for the Future — Ch. 3 Lesson Nine

Background

This concluding activity is to wrap up the lessons presented in this guide and remind students that one person can make a world of difference. Ask students to create a vision for their future and what role they will take in sustaining a healthy community for themselves, family and friends and neighbors.

Goal

Students will consider one or more actions they will commit to following and share that action with others.

Objectives

1. Create or adopt a personal slogan.
2. Write and illustrate their vision for the future.

Procedures

Personal Slogan

1. Introduce the assignment of creating a personal slogan to wear and/or share with other students, family and friends.
2. Explain what a slogan is and its purpose. Students may be able to connect with slogans by describing the slogans or images of the businesses they know (e.g. pizza, fast food, clothing)
3. Use the attached slogan suggestions as a starting point for discussion to give students ideas.
4. Set up tables with supplies to make slogans as buttons, stickers or decorated bookmark.
5. Students receive a full sheet of paper to create their bookmarks. See instruction sheet.

Vision Statement

Class discussion

1. Explain what a vision is and introduce students to creating their personal "picture" for the future.
2. Write the following quote on the board: "Thousands of tired, nerve-shaken, over-civilized people are beginning to find out that going to the mountains is going home; that wildness is a necessity; and that mountain parks and reservations are useful not only as fountains of timber and irrigating rivers, but as   fountains of life."
3. Ask students to reflect on the meaning of this quote. Does this sound like a vision statement?
4. Ask what else do they remember about John Muir, the speakers and activities that have informed the actions they can take now and into the future.
5. Write student ideas on the board to review and discuss.

Vision statement project

1. Introduce the assignment to describe in their own words and drawings (they may also clip pictures from magazines or newspapers) their vision for the future and one more actions they will take to help sustain a healthy future for all people.
2. Give students time in class to get started writing the vision statement or drawings. Students can search for pictures at home or bring in old magazines to class.
3. Students will deliver an oral presentation to class (at a later date) after all vision statements are completed.

Follow-up and Evaluation

- Does it appear that the student's behavior or attitude about creating a healthy future has been influenced by this activity or any that come before?
- Have the student's actions to date, or during this activity, demonstrated a genuine interest in creating and following through on their action steps?
- What actions are the students saying they will do?
- Does the activity appear to be meaningful to the student?
- Do students have an understanding of sustainability when it comes to planning for the future?

Advance Preparation

Secure supplies to make bookmarks or buttons for personal slogans

Supplies

- ✓ Heavy weight white paper, index card stock or construction paper in 8 1/2 x 11 sheets
- ✓ Ribbon, yarn, other decorative art supplies and white or gel glue
- ✓ One hole punch
- ✓ Markers, colored pencils

Common Core Standards

Third grade. Presentation of Knowledge and Ideas. 4, 6
Fourth grade. Presentation of Knowledge and Ideas. 4, 6

Extensions

Create bookmarks or buttons to sell as a fundraiser for an environmental project at the school.

Suggestions for Personal Slogans

Be a friend to the earth!
Keep exploring!
Ask questions and wonder.
Step lightly.
Live simple.
Care for the environment.
Plant a tree (or a seed).

Live by example.
Take Action!
Clean a place that is dirty.
Share your passion for the earth with those you love.
Make your own discoveries
Go into the wilderness.

Instructions for Bookmarks

1. Write your personal slogan on a bookmark and decorate it by drawing a picture, or by gluing on ribbons or other art supplies. You can complete the bookmarks at home that you don't finish during class.
2. Use one sheet of paper to make four bookmarks.
3. Fold paper in half across the short side and then in half again.
4. Unfold the paper and cut on the fold lines.
5. Write your name and personal slogan on one or both sides before decorating the bookmark.
6. Cut a strip of ribbon to about 9" in length
7. Use a one hole punch to punch a hole in the top of the bookmark
8. Thread the ribbon through the hole and tie the ribbon into a double knot, so the knot is tied tightly.

Glossary

Biodegradable. When an object can be decomposed either by living organisms or naturally over time back into the earth.

Biology. The branch of science that deals with the origin, history, physical characteristics, life processes and habits of living organisms.

Botany. The branch of science that studies plants, life cycle, structure, growth and classification.

Climate. Weather patterns of a particular place that occur over a period of years, such as cloudiness, temperature, air pressure, humidity, rainfall and winds.

Conservation. A complex term that addresses the careful use of natural resources to prevent injury, waste, decay or loss. Also refers to the official supervision of rivers, forests, and other natural resources in order to preserve and protect them through prudent management.

Deforestation. Cutting down (or clearing) massive amounts of trees so quickly, the forest has no time to grow back. This action often results in damage to the quality of the land.

Ecology. The branch of science that examines the relationships that organisms have to each other and to their environment. Scientists who study those relationships are called ecologists.

Ecosystem. A geographic area where plants, animals, and other organisms, as well as weather and landscape, work together and depend on each other to sustain life.

Environment. The physical world we live in, including all the circumstances that surround and influence life on earth, the atmosphere, food chain and water cycle.

Garden. The area immediately surrounding the Muir House, where trees, shrubs and flowers are planted. This is separate from the ranch used for agriculture.

Geology. The area of science that deals with the dynamics, physical history and structure of the earth; including the rocks and rock formations, and the physical, chemical, and biological changes that have occurred and continue to occur.

Glacial Moraine. A mass of loose rock, soil, and earth that sits in front of a glacier as it moves down a drainage or valley and then deposited by the edge of a glacier when movement has stopped. A terminal moraine marks the location where the glacier stopped.

Glacial Valley. A steep-sided, U-shaped valley formed by erosion when glaciers move through a drainage or river channel

Glacier. An extended mass of ice formed from snow falling and accumulating where the rate of snowfall exceeds the rate at which snow melts. The ice moves very slowly, either descending from high mountains, as in valley glaciers, or moving outward from centers of accumulation, as in continental glaciers

Hetch Hetchy. Name of the reservoir formed when a dam was built that flooded the Hetch Hetchy Valley. The existence of the dam to this day remains controversial. The Restore Hetch Hetchy group continues its fervent advocacy efforts in favor of its restoration. The dam is located within the boundary of Yosemite National Park and provides water for San Francisco residents and businesses

Horticulture: The art or science of growing flowers, fruits, vegetables and shrubs in gardens or orchards

Mt. Wanda and Mt. Helen. Mountain tops that John Muir named after his daughters.

Muir Glacier. A glacier located in Southeast Alaska, in the St. Elias Mountains, flowing southeast from Mount Fairweather. Covers about 350 square miles.

Natural history. The sciences, such as botany, mineralogy, or zoology, concerned with the study of all subjects in the natural world.

Naturalist. A person who either studies, or is an expert in, nature or natural history as a result of direct observation of animals or plants.

Curriculum Guide Bibliography

Arrowhead Pine Cone Festival. Fun pinecone facts. Retrieved May 17, 2012. http://pineconefestival.org/Fun_Pine_Cone_Facts.html

Balgooy, V. (2004, Fall). Creating award winning school programs. Forum Journal. 19(1). 31-38.

Brown, K. (1998). Environmental Service-Learning. St. Louis, MN: Tree Trust.

Browning, P. (Ed.). (1988). John Muir in his own words: A book of quotations. Lafayette, CA: Great West Books.

City of Martinez. Demographics. (n.d.) Retrieved: August 14, 2011. http://www.cityofmartinez.org/depts/econdev/demographics.asp

Corbett, J.B. (2006). Communicating Nature, How we create and understand environmental messages. Washington, DC: Island Press.

Cornell, J. (1998). Sharing nature with children. Nevada City, CA: Dawn Publications.

DeSpain, P. (1994). Twenty-two Splendid Tales to tell From Around the World. Vol. 2, 3rd ed. Little Rock, AR: August House, Inc.

Evans, A. (2002, Feb.). A crateful of American folk art. Antiques and Collecting. (107) 35-37, 63-65.

Foodreference.com. Food History Timeline 1880 to 1884. Retrieved June 12, 2012. http://www.foodreference.com/html/html/food-history-1880.html

Gilbertson, B., Bates, T., McLaughlin, T. & Ewert, A. (2006). Outdoor education: Methods and strategies. Champaign, Il: Human Kinetics.

Herman, M.L., Passineau, J.F., Schimpf, A.L. Schimpf & Trener, P. (1991). Teaching kids to love the earth. Duluth, MN: Pheifer-Hamilton.

Idaho Forest Products Commission. What Is A Healthy Forest? Retrieved May 30, 2012. http://www.idahoforests.org/health1.htm.

Kidzone. Frog fact index. Retrieved June 12, 2012. http://www.kidzone.ws/lw/frogs/facts4.htm

Killion and Davison. Cultural Landscape Report of the John Muir National Historic Site. (2005).

Lambert, T. Children in the 19th Century. Retrieved June 1, 2012. http://localhistories.org/19thcenturychildren.html

Leave No Trace Center for Outdoor Ethics. Leave No Trace Principles for Kids. Retrieved: June 6, 2012. http://lnt.org/learn/7-principles

Lehrman, F. (1990). Loving the earth, a sacred landscape for children. Berkeley, CA: Celestial Arts.

Leslie, C.W. (2010). The nature connection: An outdoor workbook for kids, families and classrooms. North Adams, MA: Storey Publishing.

Martinez Historical Society. (n.d.) The history of Martinez. Retrieved: August 15, 2011. http://www.martinezhistory.org/html/martinez_history.html

Mayntz, M. 15 fun facts about woodpeckers. Retrieved June 12, 2012. http://birding.about.com/od/birdprofiles/a/15-Fun-Facts-About Woodpeckers.htm

National Geographic. Squirrels. Retrieved May 17, 2012. http://animals.nationalgeographic.com/animals/mammals/squirrel/

National Geographic. Beaver. Retrieved May 17, 2012. http://animals.nationalgeographic.com/animals/mammals/beaver/

National Geographic Kids. Animals, Creature Features: Ladybugs. Retrieved May 17, 2012. http://kids.nationalgeographic.com/kids/animals/creaturefeature/ladybug/

Pima Community College. Rattlesnake facts. Retrieved May 17, 2012. http://wc.pima.edu/Bfiero/tucsonecol109/boxes/rattlesnake.htm

Roa, M. (2011). The Conifer Connection: A guide for learning and teaching about coniferous forests and watersheds. California State Parks. Retrieve at www.caltrees.org.

Sierra Club. (2011). John Muir Day Study Guide. Retrieved: July 4, 2011. http://www.sierraclub.org/john_muir_exhibit/educational_resources/

Sima, K. & Cordi, K. (2003). Raising voices: Creating youth storytelling groups and troupes. Westport, CT: Libraries Unlimited.

Story Arts. Story Library. Retrieved June 12, 2012. http://www.storyarts.org/org/library.

The Butterfly Website. Frequently asked questions. Retrieved May 17, 2012. http://butterflywebsite.com/faq.cfm

The History of Fruit Crate Labels and Can Labels. Retrieved June 1, 2012. http://www.thelabelman.com/history_label.php

Ward, C. W. & Wilkinson, A.E. (2006). Conducting Meaningful Interpretation: A Field Guide for Success. Golden, CO: Fulcrum Publishing.

About the Author

Janice Kelley worked collaboratively with the interpretive staff at John Muir National Historic Site to prepare this field trip curriculum experience.

Janice is an award-winning author and naturalist. She is passionate about helping individuals and families to make meaningful connections to the outdoor world, their personal histories and community stories. She is the founder and director of the Nature Detectives weekly outdoor experience program held at schools and nature areas. She has coordinated community education, outreach and training programs for nonprofit organizations, visited hundreds of classrooms as a guest artist in residence, and led nature-based interpretive tours for public agencies. Janice completed her Master of Science degree from the Department of Recreation, Parks and Tourism at California State University, Sacramento, in May 2013, and received the Faculty Award of Merit.

You may contact Janice directly at outdoorjan@att.net. To learn more about Janice's professional services, programs and products, visit her website at https://naturelegacies.com. Learn more about Nature Detectives programs, products and the nature-health-child connection at https://naturedetectivesusa.com.

Practices in Environmental Stewardship